Other Books by the Same Author

An Introduction to Social Psychology
Luce & Co., Boston

Body and Mind, a History and Defense of Animism
The Macmillan Co., N. Y.

Psychology, the Study of Behavior
The Home University Library, H. Holt & Co., N. Y.

Primer of Physiological Psychology
Dent & Co., London

Is America Safe for Democracy?
Scribners

The Pagan Tribes of Borneo
(In conjunction with Dr. C. Hose)
The Macmillan Co

The Group Mind

A Sketch of the Principles of Collective Psychology with Some Attempt to Apply Them to the Interpretation of National Life and Character

By

William McDougall

Professor of Psychology at Harvard University

“Une nation est une âme, un principe spirituel. Deux choses qui, à vrai dire, n’en font qu’une constituent cette âme, ce principe spirituel. L’une est dans le passé, l’autre dans le présent. L’une est la possession en commun d’un riche legs de souvenirs; l’autre est le consentement actuel, le désir de vivre ensemble, la volonté de continuer à faire valoir l’héritage qu’on a reçu indivis.”

Ernest Renan.

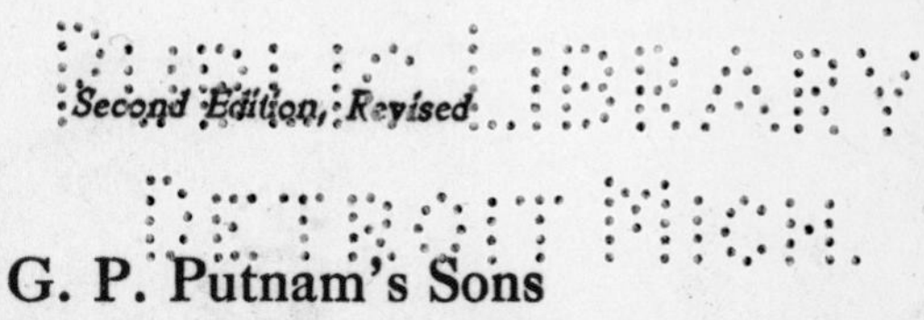

Second Edition, Revised

G. P. Putnam’s Sons
New York and London
The Knickerbocker Press

Made in the United States of America

TO

PROFESSOR L. T. HOBHOUSE

in admiration of his work in philosophy, psychology, and sociology, and in the hope that he may discern in this book some traces of the spirit by which his own writings have been inspired.

Preface to the American Edition

TO the American edition of this book I add these few words of thanks to the American readers of my previous writings; their appreciation has encouraged me to persevere in the plan of writing a Treatise on Social Psychology, of which plan the present volume represents the second step. In America the public interest in psychology and sociology is much more widespread than in these islands, a fact sufficiently attested by the existence of chairs in these subjects in all the leading universities, and the lack of such chairs in all but two or three of the universities in this country. By those Englishmen who believe that the study of these sciences is a matter of urgent national importance this state of affairs is deplored, and they desire and even hope that in this matter the example of America may soon be followed here. Meanwhile I send this book across the water, in the hope that it may contribute its mite towards the working-out of the great experiment in Social Science which the American people is making with so ardent faith in the power of the Group Mind to attain to effective direction of its own development.

W. McDOUGALL.

OXFORD, 1920.

Preface

IN this book I have sketched the principles of the mental life of groups, and have made a rough attempt to apply these principles to the understanding of the life of nations. I have had the substance of the book in the form of lecture notes for some years, but have long hesitated to publish it. I have been held back, partly by my sense of the magnitude and difficulty of the subject, and the inadequacy of my own preparation for dealing with it, partly because I wished to build upon a firm foundation of generally accepted principles of human nature.

Some fifteen years ago I projected a complete treatise on Social Psychology which would have comprised the substance of the present volume. I was prevented from carrying out the ambitious scheme, partly by the difficulty of finding a publisher, partly by my increasing sense of the lack of any generally accepted or acceptable account of the constitution of human nature. I found it necessary to attempt to provide such a foundation, and in 1908 published my *Introduction to Social Psychology*. That book has enjoyed a certain popular success. But it was more novel, more revolutionary, than I had supposed when writing it; and my hope that it would rapidly be accepted by my colleagues as, in the main, a true account of the fundamentals of human nature has not been realised.

All this part of psychology labours under the great difficulty that the worker in it cannot, like other men of science, publish his conclusions as discoveries which will necessarily be accepted by any persons competent to judge. He can only state his conclusions and his reasoning and hope that they may gradually gain the general approval of his colleagues. For to the obscure questions of fact with which he deals, it is in the nature of things impossible to return answers supported by indisputable experimental proofs. In this field the evidence of an author's approximation towards truth can consist only in his success in gradually persuading competent opinion of the value of his views. My sketch of the fundamentals of human nature can hardly claim even that degree of success which would be constituted by an active criticism and discussion of it in competent quarters. Yet there are not wanting indications that opinion is turning slowly towards the acceptance of some such doctrine as I then outlined. Especially the development of psycho-pathology, stimulated so greatly by the esoteric dogmas of the Freudian school, points in this direction. The only test and verification to which any scheme of human nature can be submitted is the application of it to practice in the elucidation of the concrete phenomena of human life, and in the control and direction of conduct, especially in the two great fields of medicine and education. And I have been much encouraged by finding that some workers in both of these fields have found my scheme of use in their practice, and have even, in some few cases, given it a cordial general approval. But group psychology is itself one of the fields in which such testing and verification must be sought. And I have decided to delay no longer in attempting to bring my scheme to this test. I am also impelled to venture on what may appear to be premature publication by the fact that five of the best years of my life have been wholly given up to military service and the practical problems of psycho-therapy, and by the reflection that the years of a man's life are numbered and that, even though I should delay yet another fifteen years, I might find that I had made but little progress towards securing the firm foundation I desired.

It may seem to some minds astonishing that I should now admit that the substance of this book was committed to writing before the Great War; for that war is supposed by some to have revolutionised all our ideas of human nature and of national life. But the war has given me little reason to add to or to change what I had written. This may be either because I am too old to learn, or because what I had written was in the main true; and I am naturally disposed to accept the second explanation.

I wish to make it clear to any would-be reader of this volume that it is a sequel to my *Introduction to Social Psychology*, that it builds upon that book and assumes that the reader is acquainted with it. That former volume has been criticised as an attempted outline of *Social Psychology*. One critic remarks that it may be good psychology, but it is very little social; another wittily says, "Mr. McDougall, while giving a full account of the genesis of instincts that act in society, hardly shows how they issue into society. He seems to do a great deal of packing in preparation for a journey on which he never starts." The last sentence exactly describes the book. I found myself, like so many of my predecessors and contemporaries, about to start on a voyage of exploration of societies with an empty trunk, or at least with one very inadequately supplied with the things essential for successful travelling. I decided to avoid the usual practice of starting without impedimenta, and of picking up or inventing bits of make-shift equipment as each emergency arose; I would pack my trunk carefully before starting. And now, although my fellow-travellers have not entirely approved my outfit, I have launched out to put it to the test; and I cannot hope that my readers will follow me if they have not at their command a similar outfit—namely, a similar view of the constitution of human nature.

I would gratefully confess that the resolve to go forward without a further long period of preparation has been made possible for me largely by the encouragement I have had from the recently published work of Dr. James Drever, *Instinct in Man*. For the author of that work has carefully studied the most fundamental part of my *Social Psychology*, in the light of his wide knowledge of the cognate literature, and has found it to be in the main acceptable.

The title and much of the substance of the present volume might lead a hasty reader to suppose that I am influenced by, or even in sympathy with, the political philosophy associated with German "idealism." I would, therefore, take this opportunity both to prevent any such erroneous inference, and to indicate my attitude towards that system of thought in plainer language than it seemed possible to use before the war. I have argued that we may properly speak of a group mind, and that each of the most developed nations of the present time may be regarded as in process of developing a group mind. This must lay me open to the suspicion of favouring the political philosophy which makes of the state a super-individual and semi-divine person before whom all men must bow down, renouncing their claims to freedom of judgment and action; the political philosophy in short of German "idealism," which derives in the main from Hegel, which has been so ably represented in this country by Dr. Bosanquet, which has exerted so great an influence at Oxford, and which in my opinion is as detrimental to honest and clear thinking as it has proved to be destructive of political morality in its native country. I am relieved of the necessity of attempting to justify these severe strictures by the recent publication of *The Meta-*

physical Theory of the State by Prof. L. T. Hobhouse. In that volume Professor Hobhouse has subjected the political philosophy of German "idealism," and especially Dr. Bosanquet's presentation of it, to a criticism which, as it seems to me, should suffice to expose the hollowness of its claims to all men for all time; and I cannot better define my own attitude towards it than by expressing the completeness of my sympathy with the searching criticism of Mr. Hobhouse's essay. In my youth I was misled into supposing that the Germans were the possessors of a peculiar wisdom; and I have spent a large part of my life in discovering, in one field of science after another, that I was mistaken. I can always read the works of some German philosophers, especially those of Hermann Lotze, with admiration and profit; but I have no longer any desire to contend with the great systems of "idealism," and I think it a cruel waste that the best years of the lives of many young men should be spent struggling with the obscure phrases in which Kant sought to express his profound and subtle thought. My first scientific effort was to find evidence in support of a new hypothesis of muscular contraction; and, in working through the various German theories, I was dismayed by their lack of clear mechanical conceptions. My next venture was in the physiology of vision, a branch of science which had become almost exclusively German. Starting with a prepossession in favour of one of the dominant German theories, I soon reached the conclusion that the two German leaders in this field, Helmholtz and Hering, with their hosts of disciples, had, in spite of much admirable detailed work, added little of value and much confusion to the theory of vision left us by a great Englishman,—namely, Thomas Young; and in a long series of papers I endeavoured to restate and supplement Young's theory. Advancing into the field of physiological psychology, I attacked the ponderous volumes of Wundt with enthusiasm; only to find that his physiology of the nervous system was a tissue of unacceptable hypotheses and that he failed to connect it in any profitable manner with his questionable psychology. And, finding even less satisfaction in such works as Ziehen's *Physiologische Psychologie*, with its crude materialism and associationism, or in the dogmatic speculations of Verworn, I published my own small attempt to bring psychology into fruitful relations with the physiology of the nervous system. This brought me up against the great problem of the relations between mind and body; and, having found that, in this sphere, German "idealism" was pragmatically indistinguishable from thorough-going materialism, and that those Germans who claimed to reconcile the two did not really rise much above the level of Ernst Haeckel's wild flounderings, I published my *History and Defense of Animism*. And in this field, though I found much to admire in the writings of Lotze, I derived most encouragement and stimulus from Professor Bergson. In working at the foundations of human nature, I found little help in German psychology, and more in French books, especially in those of Professor Ribot. In psycho-pathology I seemed to find that the claims of the German and Austrian schools were far outweighed by those of the French writers, especially of Professor Janet. So now, in attacking the problems of the mental life of societies, I have found little help from German psychology or sociology, from the elaborations of Wundt's *Völkerpsychologie* or the ponderosities of Schaeffle, and still less from the "idealist" philosophy of politics. In this field also it is French authors from whom I have learnt most, and with whom I find myself most in sympathy, especially MM. Fouillée, Boutmy, Tarde, and Demolins; though I would not be thought to hold in low esteem the works of many English and American authors, notably those of Buckle,

Bagehot, Maine, Lecky, Lowell, and of many others, to some of which I have made reference in the chapters of this book.

I have striven to make this a strictly scientific work, rather than a philosophical one; that is to say, I have tried to ascertain and state the facts and principles of social life as it is and has been, without expressing my opinion as to what it should be. But, in order further to guard myself against the implications attached by German "idealism" to the notion of a collective mind, I wish to state that politically my sympathies are with individualism and internationalism, although I have, I think, fully recognised the great and necessary part played in human life by the Group Spirit, and by that special form of it which we now call "Nationalism."

I know well that those of my readers whose sympathies are with Collectivism, Syndicalism, or Socialism in any of its various forms will detect in this book the cloven foot of individualism, and leanings towards the aristocratic principle. I know also that many others will reproach me with giving countenance to communistic and ultra-democratic tendencies. I would, therefore, point out explicitly at the outset that, if this book affords justification for any normative doctrine or ideal, it is for one which would aim at a synthesis of the principles of individualism and communism, of aristocracy and democracy, of self-realisation and of service to the community. I can best express this ideal in the wise words of Mr. F. H. Bradley, which I extract from his famous essay on "My Station and its Duties." "The individual's consciousness of himself is inseparable from the knowing himself as an organ of the whole; . . . for his nature now is not distinct from his 'artificial self.' He is related to the living moral system not as to a foreign body; his relation to it is 'too inward even for faith,' since faith implies a certain separation.

It is no other-world that he cannot see but must trust to; he feels himself in it, and it in him; . . . the belief in this real moral organism is the one solution of ethical problems. It breaks down the antithesis of despotism and individualism; it denies them, while it preserves the truth of both. The truth of individualism is saved, because, unless we have intense life and self-consciousness in the members of the state, the whole state is ossified. The truth of despotism is saved, because, unless the member realises the whole by and in himself, he fails to reach his own individuality. Considered in the main, the best communities are those which have the best men for their members, and the best men are the members of the best communities. . . . The two problems of the best man and best state are two sides, two distinguishable aspects of the one problem, how to realise in human nature the perfect unity of homogeneity and specification; and when we see that each of these without the other is unreal, then we see that (speaking in general) the welfare of the state and the welfare of its individuals are questions which it is mistaken and ruinous to separate. Personal morality and political and social institutions cannot exist apart, and (in general) the better the one, the better the other. The community is moral, because it realises personal morality; personal morality is moral, because and in so far as it realises the moral whole."

Since correcting the proofs of this volume, I have become acquainted with two recent books whose teaching is so closely in harmony with my own that I wish to direct my readers' attention to them. One is Sir Martin Conway's *The Crowd in Peace and War*, which contains many valuable illustrations of group life. The other is Miss M. P. Follett's *The New State; Group Organization the Solution of Popular Government*, which expounds the principles and advantages of collective deliberation with vigour and insight.

I am under much obligation to Prof. G. Dawes Hicks. He has read the proofs of my book, and has helped me greatly with many suggestions; but he has, of course, no responsibility for the views expressed in it.

W. McD.

OXFORD, March, 1920.

CONTENTS

The Group Mind

The Group Mind

CHAPTER I

Introduction

The Province of Collective Psychology

TO define exactly the relations of the several special sciences is a task which can never be completely achieved so long as these sciences continue to grow and change. It is a peculiarly difficult task in respect of the biological sciences, because we have not yet reached general agreement as to the fundamental conceptions which these sciences should employ. To illustrate this difficulty I need only refer to a recent symposium of the Aristotelian Society in which a number of distinguished philosophers and biologists discussed the question "Are physical, biological, and psychological categories irreducible?" The discussion revealed extreme differences of opinion, and failed to bring the disputants nearer to a common view. The difficulty is still greater in respect of human sciences—anthropology, psychology, ethics, politics, economics, sociology, and the rest; and it is not to be hoped that any general agreement on this difficult question will be reached in the near future. Yet it seems worth while that each writer who aspires to break new ground in any part of this field of inquiry should endeavour to make clear to himself and others his conception of the relations of that part to the rest of the field. It is, then, in no dogmatic spirit, or with any belief in the finality of the position assigned to my topic, that I venture the following definition of the province of psychology with which this book is concerned.

I have chosen the title, "The Group Mind," after some hesitation in favour of the alternative, "Collective Psychology." The latter has the advantage that it has already been used by several continental authors, more especially French and Italian psychologists. But the title I have chosen is, I think, more distinctively English in quality and denotes more clearly the topic that I desire to discuss.

An alternative and not inappropriate title would have been "An Outline of Social Psychology"; but two reasons prevented the adoption of this. First, my *Introduction to Social Psychology* has become generally known by the abbreviated title *Social Psychology.* This was an unforeseen result and unfortunate designation; for, as I have explained in the Preface to the present volume, that other work was designed merely as a propædeutic; it aimed merely at clearing the ground and laying the foundations for Social Psychology, while leaving the topic itself for subsequent treatment. Secondly, I conceive Group Psychology to be a part only, though a very large part, of the total field of Social Psychology; for, while the former has to deal only with the life of groups, the latter has also to describe and account for the influence of the group on the growth and activities of the individual. This is the most concrete part of psychology and naturally comes last in the order of development of the science; for, like other sciences, psychology began with the most abstract notions, the forms of activity of mind in general, and, by the aid of the abstract conceptions achieved by the earlier workers, progresses to the consideration of more concrete problems, the problems presented by actual living persons in all their inexhaustible richness and complexity.

Until the later decades of the nineteenth century, psychology continued to concern itself almost exclusively with the mind of man conceived in an abstract fashion, not as the mind of any particular individual, but as the mind of a representative individual considered in abstraction from his social settings as something given to our contemplation fully formed and complete.

Two important changes of modern thought have shown the necessity of a more concrete treatment of psychological problems. The first has been the coming into prominence of the problems of genesis which, although not originated by Darwin, received so great an impetus from his work. The second has been the increasing realisation of the need for a more synthetic treatment of all fields of science, the realisation that analysis alone carries us ever farther away from concrete problems and leads only to a system of abstract conceptions which are very remote from reality, however useful they may prove in the physical sciences. The biological and the human sciences especially have been profoundly affected by these two changes of modern thought. As Theodore Merz has so well shown in the fourth volume of his monumental work,[1] the need has been increasingly felt of the *vue d'ensemble*, of the synthetic mode of regarding organisms, men, and institutions, not as single things, self-contained and complete in themselves, but as merely nodes or meeting points of all the forces of the world acting and reacting in unlimited time and space.

Psychology was, then, until recent years the science of the abstract individual mind. Each worker aimed at

[1] *History of European Thought in the 19th Century.*

rendering by the aid of introspection an analytic description of the stream of his own consciousness, a consistent classification of the elements or features that he seemed to discover therein, and some general laws or rules of the order of succession and conjunction of these features; postulating in addition some one or more explanatory principles or active agencies such as "the will" or the desire of pleasure, the aversion from pain, or "the association of ideas," to enable him to account for the flow of the distinguishable elements of consciousness. The psychology achieved by these studies, necessary and valuable as they were, was of little help to men who were struggling with the concrete problems of human life and was therefore largely ignored by them. But, as I have pointed out in the Introduction to my *Social Psychology*, those who approached these problems were generally stimulated to do so by their interest in questions of right and wrong, in questions of norms and standards of conduct, the urgency of which demanded immediate answers for the practical guidance of human life in all its spheres of activity, for the shaping of laws, institutions, governments, and associations of every kind; or, as frequently perhaps, for the justification and defence of standards of conduct, modes of belief, and forms of institution, which men had learnt to esteem as supremely good.

Thus the political science of Hobbes was the expression of his attempt to justify the monarchy established by the Tudors and endangered by the failings of the Stuart kings; while that of Locke was equally the outcome of his desire to justify the revolution of 1688. Hobbes felt it worth while to preface his *magnum opus* on political philosophy with a fanciful sketch of human nature and of primitive society; yet, as Mr. Gooch remarks, "neither Hobbes nor his contemporaries knew anything of the actual life of primitive communities."[1] And it may be added that they knew as little of the foundations of human nature. Again, the social doctrines of Rousseau, with all their false psychology, were formulated in order to stir men to revolt against the conditions of social life then prevalent in Europe. In a similar way, in the development of all that body of social doctrine that went under the name of Utilitarianism and which culminated in the political science and economy of the Manchester School, every step was prompted by the desire to find theoretical guidance or justification for rules governing human activity. And, if we go back to the *Politics* of Aristotle, we find the normative or regulative aim still more prominent.

Thus, in all the human sciences, we see that the search for what is has been inextricably confused with and hampered by the effort to show what ought to be; and the further back we go in their history, the more does the normative point of view predominate. They all begin in the effort to describe what ought to be; and incidentally give some more or less fallacious or fantastic account of what is, merely in order to support the normative doctrines. And, as we trace their history forward towards the present time, we find the positive element coming more and more to the front, until it tends to preponderate over and even completely to supplant the normative aim. Thus even in Ethics there is now perceptible in some quarters a tendency to repudiate the normative standpoint. All the social sciences have, then, begun their work at what, from the strictly logical point of view, was the wrong end; instead of first securing a basis of positive science and then building up the normative doctrines upon that basis, they have advanced by repeatedly going backwards towards what should have been their foundations. Now the most

[1] *Political Thought in England from Bacon to Halifax*, Home University Library, p. 49.

important part of the positive basis of the social sciences is psychology; we find accordingly the social sciences at first ignoring psychology and then gradually working back to it; they became gradually more psychological and, in proportion as they did so, they became more valuable. Modern writers on these topics fall into two classes; those who have attempted to work upon a psychological foundation, and those who have ignored or denied the need of any such basis. The earlier efforts of the former kind, among which we may reckon those of Adam Smith, Bentham, and the Mills, although they greatly influenced legislation and practice in general, have nevertheless brought the psychological method into some disrepute, because they reasoned from psychological principles which were unduly simplified and in fact misleading, notably the famous principle of psychological hedonism on which they so greatly relied. Their psychology was, in brief, too abstract; it had not achieved the necessary concreteness, which only the introduction of the genetic standpoint and the *vue d'ensemble* could give it. Other writers on the social sciences were content to ignore the achievements of psychology; but, since they dealt with the activities of human beings and the products of those activities, such as laws, institutions and customs, they could hardly avoid all reference to the human mind and its processes; they then relied upon the crude unanalysed psychological conceptions of popular speech; often they went further and aspiring to explain the phenomena they described, made vast assumptions about the constitution and working of the human mind. Thus, for example, Renan, when he sought to explain some feature of the history of a nation or society, was in the habit, like many others, of ascribing it to some peculiar instinct which he postulated for this particular purpose, such as a political or a religious instinct or an instinct of subordination or of organisation. Comte made egoism and altruism the two master forces of the mind. Sir Henry Maine asserted that "satisfaction and impatience are the two great sources of political conduct," and, after asserting that "no force acting on mankind has been less carefully examined than Party, and yet none better deserves examination," he was content to conclude that "Party is probably nothing more than a survival and a consequence of the primitive combativeness of mankind."[1] More recently Prof. Giddings has discovered the principal force underlying all human associations in *Consciousness of Kind*. Butler and the intuitive moralists postulated "conscience" or moral sense as something innately present in the souls of men; while the creators of the classical school of political economy were for the most part content to assume that man is a purely rational being who always intelligently pursues his own best interest, a false premise from which they deduced some conclusions that have not withstood the test of time. Similar vague assumptions may be found in almost every work on the social sciences—all illustrating the need for a psychology more concrete than the older individual psychology, as a basis for these sciences, a positive science, not of some hypothetical Robinson Crusoe, but of the mental life of men as it actually unfolds itself in the families, tribes, nations, societies of all sorts, that make up the human world.

The general growth of interest in genetic problems, stimulated so greatly by the work of Darwin, turned the attention of psychologists to the problem of the genesis of the developed human mind—the problem of its evolution in the race and its development in the individual. Then it at once became apparent that both these processes are essentially social; that they involve, and at every step are determined by, interactions between the individual and

[1] Essay on "The Nature of Democracy" in *Popular Government*, London, 1885.

his social environment; that, while the growth of the individual mind is moulded by the mental forces of the society in which it grows up, those forces are in turn the products of the interplay of the minds composing the society; that, therefore, we can only understand the life of individuals and the life of societies, if we consider them always in relation to one another. It was realised that each man is an individual only in an imcomplete sense; that he is but a unit in a vast system of vital and spiritual forces which, expressing themselves in the form of human societies, are working towards ends which no man can foresee; a unit whose chief function it is to transmit these forces unimpaired, which can change or add to them only in infinitesimal degree, and which, therefore, has but little significance and cannot be accounted for when considered in abstraction from that system. It became clear that the play of this system of forces at any moment of history is predominantly determined by conditions which are themselves the products of an immensely long course of evolution, conditions which have been produced by the mental activities of countless generations and which are but very little modified by the members of society living at any one time; so that, as has been said, society consists of the dead as well as of the living, and the part of the living in determining its life is but insignificant as compared with the part of the dead.

Any psychology that recognises these facts and attempts to display the reciprocal influences of the individual and the society in which he plays his part may be called Social Psychology. Collective or Group Psychology is, then, a part of this larger field. It has to study the mental life of societies of all kinds; and such understanding of the group life as it can achieve has then to be used by Social Psychology in rendering more concrete and complete our understanding of the individual life.

Group Psychology itself consists properly of two parts, that which is concerned to discover the most general principles of group life, and that which applies these principles to the study of particular kinds and examples of group life. The former is logically prior to the second; though in practice it is hardly possible to keep them wholly apart. The present volume is concerned chiefly with the former branch. Only when the general principles of group life have been applied to the understanding of particular societies, of nations and the manifold system of groups within the nation, will it be possible for Social Psychology to return upon the individual life and give of it an adequate account in all its concrete fulness.

The nature of Group Psychology may be illustrated by reference to Herbert Spencer's conception of sociology. Spencer pointed out that, if you set out to build a stable pile of solid bodies of a certain shape, the kind of structure resulting is determined by the shapes and properties of these units, that for example, if the units are spheres, there are only very few stable forms which the pile can assume. The same is true, he said, of such physical processes as crystallisation; the form and properties of the whole or aggregate are determined by the properties of the units. He maintained with less plausibility that the same holds good of animal and vegetable forms and of the elements of which they are composed. And he went on to argue that, in like manner, the structure and properties of a society are determined by the properties of the units, the individual human beings, of which it is composed.

This last proposition is true in a very partial sense only. For the aggregate which is a society has, in virtue of its past history, positive qualities which it does not derive from the units which compose it at any one time; and in virtue of these qualities it acts upon its units in a manner very different from that in which the units as such interact with one another. Further, each unit, when it becomes a member of a group, displays properties or modes of reaction which it does not display, which remain latent or potential only, so long as it remains outside that group. It is possible, therefore, to discover these potentialities of the units only by studying them as elements in the life of the whole. That is to say, the aggregate which is a society has a certain individuality, is a true whole which in great measure determines the nature and the modes of activity of its parts; it is an organic whole. The society has a mental life which is not the mere sum of the mental lives of its units existing as independent units; and a complete knowledge of the units, if and in so far as they could be known as isolated units, would not enable us to deduce the nature of the life of the whole, in the way that is implied by Spencer's analogies.

Since, then, the social aggregate has a collective mental life, which is not merely the sum of the mental lives of its units, it may be contended that a society not only enjoys a collective mental life but also has a collective mind or, as some prefer to say, a collective soul.

The tasks of Group Psychology are, then, to examine the conception of the collective or group mind, in order to determine whether and in what sense this is a valid conception; to display the general principles of collective mental life which are incapable of being deduced from the laws of the mental life of isolated individuals; to distinguish the principal types of collective mental life or group mind; to describe the peculiarities of those types and as far as possible to account for them. More shortly, Group Psychology has, first, to establish the general principles of group life (this is general collective psychology); secondly, it has to apply these principles in the endeavour to understand particular examples of group life. Group Psychology, thus conceived, meets at the outset a difficulty which stands in the way of every attempt of psychology to leave the narrow field of highly abstract individual psychology. It finds the ground already staked out and occupied by the representatives of another science, who are inclined to resent its intrusion as an encroachment on their rights. The science which claims to have occupied the field of Group Psychology is Sociology; and it is of some importance that the claims of these sciences should be reconciled, so that they may live and work harmoniously together. I have no desire to claim for Group Psychology the whole province of Sociology. As I conceive it, that province is much wider than that of Group Psychology. Sociology is essentially a science which has to take a comprehensive and synthetic view of the life of mankind, and has to accept and make use of the conclusions of many other more special sciences of which psychology, and especially Group Psychology, is for it perhaps the most important. But other special sciences have very important if less intimate contributions to make to it. Thus, if it be true that great civilisations have decayed owing to changes of climate of their habitats, or owing to the introduction of such diseases as malaria into them, then Climatology and Epidemiology have their contributions to make to Sociology. If peculiarities of diet or the crossing of racial stocks may profoundly affect the vigour of peoples, Physiology must have its say. General Biology and the science of Genetics are bringing to light much that must be incorporated in Sociology. Economics, although needing to be treated far more psychologically than it commonly has been, has its special contribution to make. These are only a few illustrations of the fact that the field of Sociology is very much wider and more general than that of Group Psychology, however important to it the conclusions of the narrower science may be.

In this book it will be maintained that the conception of

a group mind is useful and therefore valid; and, since this notion has already excited some opposition and criticism and is one that requires very careful definition, some attempt to define and justify it may usefully be made at the outset; though the completer justification is the substance of the whole book. Some writers have assumed the reality of what is called the "collective consciousness" of a society, meaning thereby a unitary consciousness of the society over and above that of the individuals comprised within it. This conception is examined in Chapter II and provisionally rejected. But it is maintained that a society, when it enjoys a long life and becomes highly organised, acquires a structure and qualities which are largely independent of the qualities of the individuals who enter into its composition and take part for a brief time in its life. It becomes an organised system of forces which has a life of its own, tendencies of its own, a power of moulding all its component individuals, and a power of perpetuating itself as a self-identical system, subject only to slow and gradual change.

In an earlier work, in which I have sketched in outline the program of psychology,[1] I wrote: "When the student of behaviour has learnt from the various departments of psychology . . . all that they can teach him of the structure, genesis, and modes of operation of the individual mind, a large field still awaits his exploration. If we put aside as unproven such speculations as that touched on at the end of the foregoing chapter (the view of James that the human mind can enter into an actual union or communion with the divine mind) and refuse to admit any modes of communication or influence between minds other than through the normal channels of sense-perception and bodily movement, we must nevertheless recognise the

[1] *Psychology, the Study of Behaviour*, Home University Library, London, 1912.

existence in a certain sense of over-individual or collective minds. We may fairly define a mind as an organised system of mental or purposive forces; and, in the sense so defined every highly organised human society may properly be said to possess a collective mind. For the collective actions which constitute the history of any such society are conditioned by an organisation which can only be described in terms of mind, and which yet is not comprised within the mind of any individual; the society is rather constituted by the system of relations obtaining between the individual minds which are its units of composition. Under any given circumstances the actions of the society are, or may be, very different from the mere sum of the actions with which its several members would react to the situation in the absence of the system of relations which render them a society; or, in other words, the thinking and acting of each man, in so far as he thinks and acts as a member of a society, are very different from his thinking and acting as an isolated individual."

This passage has been cited by the author of a notable work on Sociology,[1] and made by him the text of a polemic against the conception of the group mind. He writes: "This passage contains two arguments in favour of the hypothesis of super-individual 'collective' minds, neither of which can stand examination. The 'definition' of a mind as 'an organised system of mental or purposive forces' is totally inadequate. When we speak of the mind of an individual we mean something more than this. The mind of each of us has a unity other than that of such a system." But I doubt whether Mr. Maciver could explain exactly what kind of unity it is that he postulates. Is it the unity of soul substance? I have myself contended at some length that this is a necessary postulate or hypothesis,[2]

[1] *Community*, by R. M. Maciver, London, 1917.
[2] In *Body and Mind*, London, 1911.

but I do not suppose that Maciver accepts or intends to refer to this conception. Is it the unity of consciousness or of self-consciousness? Then the answer is that this unity is by no means a general and established function of the individual mind; modern studies of the disintegration of personality have shown this to be a questionable assumption, undermined by the many facts of normal and abnormal psychology best resumed under Dr. Morton Prince's term "co-consciousness."

The individual mind is a system of purposive forces, but the system is by no means always a harmonious system; it is but too apt to be the scene of fierce conflicts which sometimes (in the graver psychoneuroses) result in the rupture and disintegration of the system. I do not know how otherwise we are to describe the individual mind than as a system of mental forces; and, until Maciver succeeds in showing in what other sense he conceives it to have "a unity other than that of such a system," his objection cannot be seriously entertained. He asks, of the alleged collective mind: "Does the system so created think and will and feel and act?"[1] My answer, as set out in the following pages, is that it does all of these things. He asks further: "If a number of minds construct by their interactivity an organisation 'which can only be described in terms of mind,' must we ascribe to the construction the very nature of the forces which constructed it?" To this I reply—my point is that the individual minds which enter into the structure of the group mind at any moment of its life do not construct it; rather, as they come to reflective self-consciousness, they find themselves already members of the system, moulded by it, sharing in its activities, influenced by it at every moment in every thought and feeling and action in ways which they can neither fully understand nor escape from, struggle as they may to free

[1] *Op. cit.*, p. 76.

themselves from its infinitely subtle and multitudinous forces. And this system, as Maciver himself forcibly insists in another connection, does not consist of relations that exist external to and independent of the things related, namely the minds of individuals; it consists of the same stuff as the individual minds, its threads and parts lie within these minds; but the parts in the several individual minds reciprocally imply and complement one another and together make up the system which consists wholly of them; and therefore, as I wrote, they can "only be described in terms of mind." Any society is literally a more or less organised mental system; the stuff of which it consists is mental stuff; the forces that operate within it are mental forces. Maciver argues further: "Social organisations occur of every kind and every degree of universality. If England has a collective mind, why not Birmingham and why not each of its wards? If a nation has a collective mind, so also have a church and a trade union. And we shall have collective minds that are parts of greater collective minds, and collective minds that intersect other collective minds." By this my withers are quite unwrung. What degree of organisation is necessary before a society can properly be said to enjoy collective mental life or have a group mind is a question of degree; and the exponent of the group mind is under no obligation to return a precise answer to this question. My contention is that the most highly organised groups display collective mental life in a way which justifies the conception of the group mind, and that we shall be helped to understand collective life in these most complex and difficult forms by studying it in the simpler less elaborated groups where the conception of a group mind is less clearly applicable. As regards the overlapping and intersection of groups and the consequent difficulty of assigning the limits of groups whose unity is implied by the term group mind, I would point out that this difficulty arises only in connection with the lower forms of group life and that a parallel difficulty is presented by the lower forms of animal life. Is Maciver acquainted with the organisation of a sponge, or of the so-called coral "insect," or with that of the Portuguese man-o'-war? Would he deny the unity of a human being, or refuse to acknowledge his possession of a mind, because in these lower organisms the limits of the unit are hard or impossible to assign? Maciver goes on: "The second argument is an obvious fallacy. If each man thinks and acts differently as a member of a crowd or association and as an individual standing out of any such immediate relation to his fellows, it is still each who thinks and acts; the new determinations are determinations still of individual minds as they are influenced by aggregation. . . . But this is merely an extreme instance of the obvious fact that every mind is influenced by every kind of environment. To posit a super-individual mind because individual minds are altered by their relations to one another (as indeed they are altered by their relations to physical conditions) is surely gratuitous."[1] To this I reply—the environment which influences the individual in his life as a member of an organised group is neither the sum of his fellow members as individuals, nor is it something that has other than a mental existence. It is the organised group as such, which exists only or chiefly in the persons of those composing it, but which does not exist in the mind of any one of them, and which operates upon each so powerfully just because it is something indefinitely greater, more powerful, more comprehensive than the mere sum of those individuals. Maciver feels that "it is important to clear out of the way this misleading doctrine of super-individual minds corresponding to social or communal organisations and activities," and therefore goes on to say that "there is no more a great 'col-

[1] *Op. cit.*, p. 77.

lective' mind beyond the individual minds in society than there is a great 'collective' tree beyond all the individual trees in nature. A collection of trees is a wood, and that we can study as a unity; so an aggregation of men is a society, a much more determinate unity; but a collection of trees is not a collective tree, and neither is a collection of persons or minds a collective person or mind. We can speak of qualities of tree in abstraction from any particular tree, and we can speak of qualities of mind as such, or of some particular kind of mind in relation to some type of situation. Yet, in so doing, we are simply considering the characteristic of like elements of individual minds, as we might consider the characteristic or like elements discoverable in individual trees and kinds of trees. To conceive because of these identities, a 'collective' mind as existing *beside* those of individuals or a collective tree beside the variant examples is to run against the wall of the Idea theory." Now, I am not proposing to commit myself to this last-named theory. It is not because minds have much in common with one another that I speak of the collective mind, but because the group as such is more than the sum of the individuals, has its own life proceeding according to laws of group life, which are not the laws of individual life, and because its peculiar group life reacts upon and profoundly modifies the lives of the individuals. I would not call a forest a collective tree; but I would maintain that in certain respects a forest, a wood, or a copse, has in a rudimentary way a collective life. Thus, the forest remains the same forest though, after a hundred or a thousand years, all its constituent trees may be different individuals; and again the forest as a whole may and does modify the life of each tree, as by attracting moisture, protecting from violent and cold winds, harbouring various plants and animals which affect the trees, and so on.

But I will cite an eloquent passage from a recent work on sociology in support of my view. "The bonds of society are in the members of society, and not outside them. It is the memories, traditions, and beliefs of each which make up the social memories, traditions, and beliefs. Society like the kingdom of God is within us. Within us, within each of us, and yet greater than the thoughts and understandings of any of us. For the social thoughts and feelings and willings of each, the socialised mind of each, with the complex scheme of his relation to the social world, is no mere reproduction of the social thoughts and feelings and willings of the rest. Unity and difference here too weave their eternal web, the greater social scheme which none of us who are part of it can ever see in its entirety, but whose infinite subtlety and harmony we may more and more comprehend and admire. As a community grows in civilisation and culture, its traditions are no longer clear and definite ways of thinking, its usages are no longer uniform, its spirit is no longer to be summed up in a few phrases. But the spirit and tradition of a people become no less real in becoming more complex. Each member no longer embodies the whole tradition, but it is because each embodies some part of a greater tradition to which the freely-working individuality of each contributes. In this sense the spirit of a people, though existing only in the individual members, more and more surpasses the measure of any individual mind. Again, the social tradition is expressed through institutions and records more permanent than the short-lived members of a community. These institutions and records are as it were stored social values (just as, in particular, books may be called stored social knowledge), *in themselves nothing*, no part of the social mind, but the instruments of the communication of traditions from member to member, as also from the dead past to the living present. In this way too, with the increase of these stored values, of which members realise parts but none the whole, the spirit of a people more and more surpasses the measure of any individual mind. It is these social forces within and without, working in the minds of individuals whose own social inheritance is an essential part of their individuality, stored in the institutions which they maintain from the past or establish in the present, that mould the communal spirit of the successive generations. In this sense too a community may be called greater than its members who exist at any one time, since the community itself marches out of the past into the present, and its members at any time are part of a great succession, themselves first moulded by communal forces before they become, so moulded, the active determinants of its future moulding." An admirable statement! "The greater social scheme which none of us can see in its entirety"—"the spirit of a people" which "more and more surpasses the measure of any individual mind"—"the communal spirit of the successive generations"—"the community" which is "greater than its members who exist at any one time"; all these are alternative designations of that organised system of mental forces which exists over and above, though not independently of, the individuals in each of whom some fragment of it is embodied and which is the group mind. And the writer of this statement is Mr. R. M. Maciver; the passage occurs in the section of his book designed to "clear out of the way this misleading doctrine of super-individual minds." In the same section he goes on to say that "every association, every organised group, may and does have rights and obligations which are not the rights and obligations of any or all of its members taken distributively but only of the association acting as an organised unity. . . . As a unity the association may become a 'juristic person,' a 'corporation,' and from the legal standpoint the character of unity so conceived is very important. . . . The 'juristic person' is a real *unity*, and therefore more than a *persona ficta*, but the reality it possesses is of a totally different order of being from that of the persons who establish it." But, perversely as it seems to me, Maciver adds, "the unity of which we are thinking is not mechanic or organic or even psychic." I cannot but think that, in thus denying the organic and psychic nature of this unity, Maciver is under the influence of that unfortunate and still prevalent way of thinking of the psychic as identical with the conscious which has given endless trouble in psychology; because it has prompted the hopeless attempt, constantly renewed, to describe the structure and organisation of the mind in terms of conscious stuff, ignoring the all-important distinction between mental activity, which is sometimes, though perhaps not always consciousness, and mental structure which is not. The structure and organisation of the spirit of the community is in every respect as purely mental or psychic as is the structure and organisation of the individual mind.

Maciver very properly goes on to bring his conclusions to the pragmatic test, the test of practical results. He writes: "These false analogies . . . are the sources of that most misleading antithesis which we draw between the individual and society, as though society were somehow other than its individuals. . . . Analyse these misleading analogies, and in the revelation of their falsity there is revealed also the falsity of this essential opposition of individual and society. Properly understood, the interests of 'the individual' are the interests of society."[1] But is it true that the interests of the individual are identical with the interests of society? Obviously not. We have only to think of the condemned criminal; of the mentally defective to whom every enlightened society should deny the right of procreation; of the young soldier who sacrifices his health, his limbs, his eyesight, or his life, and perhaps the welfare of his loved ones, in serving his

[1] *Op. cit.*, p. 90.

country. It is true that the progress of society is essentially an approximation towards an ideal state in which this identification would be completed; but that is an ideal which can never be absolutely realised. Nor is it even true that the interests of society are identical with the interests of the majority of its members existing at any one time. It is, I think, highly probable that, if any great modern nation should unanimously and whole-heartedly embark upon a thorough-going scheme of state-socialism, the interests of the vast majority of individuals would be greatly promoted; they would be enabled to live more prosperously and comfortably with greater leisure and opportunity for the higher forms of activity. It is, however, equally probable that the higher interests of the nation would be gravely endangered, that it would enter upon a period of increasing stagnation and diminishing vitality and, after a few generations had passed away, would have slipped far down the slope which has led all great societies of the past to destruction.

The question may be considered in relation to the German nation. As will be pointed out in a later chapter, the structure of that nation was, before the Great War, a menace to European civilisation. If the Germans had succeeded in their aims and had conquered Europe or the world, their individual interests would have been vastly promoted; they would have enjoyed immense material prosperity and a proud consciousness of having been chosen by God to rule the rest of mankind for their good. And this would have confirmed the nation in all its vices and would have finally crushed out of it all its potentialities for developing into a well-organised nation of the higher type, fitted to play an honourable part in the future evolution of mankind. The same truth appears if we consider the problem of the responsibility of the German nation for the War. So long as that people might retain its former organisation, which, I repeat, rendered it a menace to the civilisation and culture of the whole world, its antagonists could only treat it as a criminal and an outlaw to be repressed at all costs and punished and kept down with the utmost severity. But, if it should achieve a new organisation, one which will give preponderance to the better and saner elements and traditions still preserved within it, then, although it will consist of the same individuals in the main, it will have become a new or at least a transformed nation, one with which the other nations could enter into moral relations of amity or at least of mutual toleration, one which could be admitted to a place in the greater society which the League of Nations is to become. In other words, the same population would in virtue of a changed organisation, have become a different nation.

Although Maciver, in making his attack upon the conception of the group mind, has done me the honour to choose me as its exponent, I do not stand alone in maintaining it. I am a little shy of citing in its support the philosophers of the school of German "idealism," because, as I have indicated in the Preface, I have little sympathy with that school. Yet, though one may disapprove of the methods and of most of the conclusions of a school of thought, one may still adduce in support of one's opinion such of its principles as seem to be well founded. I may, then, remind the reader that the conception of the State as a super-individual, a superhuman quasi-divine personality, is the central conception of the political philosophy of German "idealism." That conception has, no doubt, played a considerable part in bringing upon Europe its present disaster. It was an instance of one of those philosophical ideas which claim to be the product of pure reason, yet in reality are adopted for the purpose of justifying and furthering some already existing interest or institution. In this case the institution in question was the Prussian state and those, Hegel and the rest, who set up this doctrine were servants of that state. They made of their doctrine an instrument for the suppression of individuality which greatly aided in producing the servile condition of the German people. Yet the distortions and exaggerations of the political philosophy of German "idealism" should not prejudice us against the germ of truth which it contains; and the more enlightened British disciples of this school, from T. H. Green onwards, have sought with much success to winnow the grain from the chaff of the doctrine; and I cannot adduce better support for the conception of the group mind than the sentences in which a recent English writer, a sympathetic student of German "idealism," sums up the results of this winnowing process.[1] Discussing the deficiencies of the individualist philosophy of the English utilitarian school, he writes: "Not a modification of the old Benthamite premises, but a new philosophy was needed; and that philosophy was provided by the idealist school, of which Green is the greatest representative. That school drew its inspiration immediately from Kant and Hegel, and ultimately from the old Greek philosophy of the city-state. The vital relation between the life of the individual and the life of the community, which alone gives the individual worth and significance, because it alone gives him the power of full moral development; the dependence of the individual, for all his rights and for all his liberty, on his membership of the community; the correlative duty of the community to guarantee to the individual all his rights (in other words, all the conditions necessary for his, and therefore for its own, full moral development)—these were the premises of the new philosophy. That philosophy could satisfy

[1] E. Barker, *Political Thought in England from Herbert Spencer to the Present Day*, Home University Library, London, 1915.

the new needs of social progress, because it refused to worship a supposed individual liberty which was proving destructive of the real liberty of the vast majority, and preferred to emphasise the moral well-being and betterment of the whole community, and to conceive of each of its members as attaining his own well-being and betterment in and through the community. Herein lay, or seemed to lie, a revolution of ideas. Instead of starting from a central individual, to whom the social system is supposed to be adjusted, the idealist starts from a central social system, in which the individual must find his appointed orbit of duty. But after all the revolution is only a restoration; and what is restored is simply the *Republic* of Plato."[1] The same writer reminds us that "both Plato and Hegel thus imply the idea of a moral organism"; and he adds, "It is this conception of a moral organism which Bradley urges. It is implied in daily experience, and it is the only explanation of that experience. 'In fact, what we call an individual man is what he is because of and by virtue of community, and communities are not mere names, but something real.' Already at birth the child is what he is in virtue of communities; he has something of the family character, something of the national character, something of the civilised character which comes from human society. As he grows, the community in which he lives pours itself into his being in the language he learns and the social atmosphere he breathes, so that the content of his being implies in its every fibre relations of community. He is what he is by including in his essence the relations of the social State. . . . And regarding the State as a system, in which many spheres (the family, for instance) are subordinated to one sphere, and all the particular actions of individuals are subordinated to their various spheres, we may call it a moral

[1] *Op. cit.*, p. 11.

organism, a systematic whole informed by a common purpose or function. As such it has an outer side—a body of institutions; it has an inner side—a soul or spirit which sustains that body. And since it is a moral organism—since, that is to say, its parts are themselves conscious moral agents—that spirit resides in those parts and lives in their consciousness. In such an organism—and this is where it differs from an animal organism, and why we have to use the word moral—the parts are conscious: they know themselves in their position as parts of the whole, and they therefore know the whole of which they are parts. So far as they have such knowledge, and a will based upon it, so far is the moral organism self-conscious and self-willing. . . . Thus, on the one hand, we must recognise that the State lives; that there is a nation's soul, self-conscious in its citizens; and that to each citizen this living soul assigns his field of accomplishment."[1] On a later page of the same book we read— "All the institutions of a country, so far as they are effective, are not only products of thought and creations of mind: they *are* thought, and they *are* mind. Otherwise we have a building without a tenant, and a body without a mind. An Oxford college is not a group of buildings, though common speech gives that name to such a group: it is a group of men. But it is not a group of men in the sense of a group of bodies in propinquity: it is a group of men in the sense of a group of minds. That group of minds, in virtue of the common substance of an uniting idea, is itself a group-mind. There is no group-mind existing apart from the minds of the members of the group; the group-mind only exists in the minds of its members. But nevertheless it exists. There is a college mind, just as there is a Trade Union mind, or even a 'public mind' of the whole community; and we are all conscious of such a mind as some-

[1] *Op. cit.*, pp. 62–64.

thing that exists in and along with the separate minds of the members, and over and above any sum of those minds created by mere addition."[1]

The political philosophers of the idealist school have not stood alone in recognising the reality of the group mind. Some of the lawyers, notably Maitland, have arrived at a very similar doctrine; and I cannot better summarise their conclusions than Barker has done in the following passage in the book from which I have already cited so freely. "The new doctrine," he writes, "runs somewhat as follows. No permanent group, permanently organised for a durable object, can be regarded as a mere sum of persons, whose union, to have any rights or duties, must receive a legal confirmation. Permanent groups are themselves persons, group-persons, with a group-will of their own and a permanent character of their own; and they have become group-persons of themselves, without any creative act of the State. In a word, group-persons are real persons; and just because they are so, and possess such attributes of persons as will and character, they cannot have been made by the State."[2]

I am not alone, then, in postulating the reality of the group mind. And I am glad to be able to cite evidence of this, because I know well that very many readers may at first find themselves repelled by this notion of a group mind, and that some of them will incline to regard it as the fantastic fad of an academic crank.

I would say at once that the crucial point of difference between my own view of the group mind and that of the

[1] *Op. cit.*, p. 74. I consider Mr. Barker's brief statement of the nature of the group mind entirely acceptable, and it has given me great pleasure to find myself in such close harmony with it. It will perhaps give further weight to the fact of our agreement, if I add that the whole of this book, including the rest of this introductory chapter, was written before I took up Mr. Barker's brilliant little volume.

[2] *Op. cit.*, p. 175.

German "idealist" school (at least in its more extreme representatives) is that I repudiate, provisionally at least, as an unverifiable hypothesis the conception of a collective or super-individual consciousness, somehow comprising the consciousness of the individuals composing the group. I have examined this conception in the following chapter and have stated my grounds for rejecting it. The difference of practical conclusions arising from this difference of theory must obviously be very great.

Several books dealing with collective psychology have been published in recent years. Of these perhaps the most notable are G. le Bon's *Psychology of the Crowd*, his *Evolution psychologique des peuples;* Sighele's *La foule criminelle;* the *Psychologie collective* of Dr. A. A. Marie; and Alfred Fouillée's *La Science sociale contemporaine.* It is noteworthy that, with the exception of the last, all these books deal only with crowds or groups of low organisation; and their authors, like almost all others who have touched on this subject, are concerned chiefly to point out how participation in the group life degrades the individual, how the group feels and thinks and acts on a much lower plane than the average plane of the individuals who compose it.

On the other hand, many writers have insisted on the fact that it is only by participation in the life of society that any man can realise his higher potentialities; that society has ideals and aims and traditions loftier than any principles of conduct the individual can form for himself unaided; and that only by the further evolution of organised society can mankind be raised to higher levels; just as in the past it has been only through the development of organised society that the life of man has ceased to deserve the epithets "nasty, brutish, and short" which Hobbes applied to it.

We seem then to stand before a paradox. Participation in group life degrades the individual, assimilating his mental processes to those of the crowd, whose brutality, inconstancy, and unreasoning impulsiveness have been the theme of many writers; yet only by participation in group life does man become fully man, only so does he rise above the level of the savage.

The resolution of this paradox is the essential theme of this book. It examines and fully recognises the mental and moral defects of the crowd and its degrading effects upon all those who are caught up in it and carried away by the contagion of its reckless spirit. It then goes on to show how organisation of the group may, and generally does in large measure, counteract these degrading tendencies; and how the better kinds of organisation render group life the great ennobling influence by aid of which alone man rises a little above the animals and may even aspire to fellowship with the angels.

PART I

GENERAL PRINCIPLES OF COLLECTIVE PSYCHOLOGY

CHAPTER II

The Mental Life of the Crowd

IT is a notorious fact that, when a number of men think and feel and act together, the mental operations and the actions of each member of the group are apt to be very different from those he would achieve if he faced the situation as an isolated individual. Hence, though we may know each member of a group so intimately that we can, with some confidence, foretell his actions under given circumstances, we cannot foretell the behaviour of the group from our knowledge of the individuals alone. If we would understand and be able to predict the behaviour of the group, we must study the way in which the mental processes of its members are modified in virtue of their membership. That is to say, we must study the interactions between the members of the group and also those between the group as a whole and each member. We must examine also the forms of group organisation and their influence upon the life of the group.

Groups differ greatly from one another in respect of the kind and degree of organisation they possess. In the simplest case the group has no organisation. In some cases the relations of the constituent individuals to one another and to the whole group are not in any way determined or fixed by previous events; such a group constitutes merely a mob. In other groups the individuals have certain determinate relations to one another which have arisen in one or more of three ways:

(1) Certain relations may have been established between the individuals, before they came together to form a group; for example, a parish council or a political meeting may be formed by persons belonging to various definitely recognised classes, and their previously recognised relations will continue to play a part in determining the collective deliberations and actions of the group; they will constitute an incipient organisation.

(2) If any group enjoys continuity of existence, certain more or less constant relations, of subordination, deference, leadership and so forth, will inevitably become established between the individuals of which it is composed; and, of course, such relations will usually be deliberately established and maintained by any group that is united by a common purpose, in order that its efficiency may be promoted.

(3) The group may have a continued existence and a more or less elaborate and definite organisation independent of the individuals of which it is composed; in such a case the individuals may change while the formal organisation of the group persists; each person who enters it being received into some more or less well-defined and generally recognised position within the group, which formal position determines in great measure the nature of his relations to other members of the group and to the group as a whole.

We can hardly imagine any concourse of human beings, however fortuitous it may be, utterly devoid of the rudiments of organisation of one or other of these three kinds; nevertheless, in many a fortuitous concourse the influence of such rudimentary organisation is so light as to be negligible. Such a group is an unorganised crowd or mob. The unorganised crowd presents many of the fundamental phenomena of collective psychology in relative simplicity; whereas the higher the degree of organisation of a group, the more complicated is its psychology. We shall, therefore, study first the mental peculiarities of the unorganised crowd, and shall then go on to consider the modifications resulting from a simple and definite type of organisation.

Not every mass of human beings gathered together in one place within sight and sound of one another constitutes a crowd in the psychological sense of the word. There is a dense gathering of several hundred individuals at the Mansion House Crossing at noon of every week-day; but ordinarily each of them is bent upon his own task, pursues his own ends, paying little or no regard to those about him. But let a fire-engine come galloping through the throng of traffic, or the Lord Mayor's state coach arrive, and instantly the concourse assumes in some degree the character of a psychological crowd. All eyes are turned upon the fire-engine or coach; the attention of all is directed to the same object; all experience in some degree the same emotion; and the state of mind of each person is in some degree affected by the mental processes of all those about him. Those are the fundamental conditions of collective mental life. In its more developed forms, an awareness of the crowd or group as such in the mind of each member plays an important part; but this is not an essential condition of its simpler manifestations. The essential conditions of collective mental action are, then, a common object of mental activity, a common mode of feeling in regard to it, and some degree of reciprocal influence between the members of the group. It follows that not every aggregation of individuals is capable of becoming a psychological crowd and of enjoying a collective life. For the individuals must be capable of being interested in the same objects and of being affected in a similar way by them; there must be a certain degree of similarity of mental constitution among the individuals, a certain mental homogeneity of the group. Let a man stand on a tub in the midst of a gathering of a hundred Englishmen and proceed to denounce and abuse England; those individuals at once become a crowd. Whereas, if the hundred men were of as many races and nations, their attention would hardly be attracted by the orator; for they would have no common interest in the topic of his discourse. Or let the man on the tub denounce the establishment of the Church of England, and the hundred Englishmen do not become a crowd; for, although all may be interested and attentive, the words of the orator evoke in them very diverse feelings and emotions, the sentiments they entertain for the Church of England being diverse in character.

There must, then, be some degree of similarity of mental constitution, of interest and sentiment, among the persons who form a crowd, a certain degree of mental homogeneity of the group. And the higher the degree of this mental homogeneity of any gathering of men, the more readily do they form a psychological crowd and the more striking and intense are the manifestations of collective life. All gatherings of men that are not purely fortuitous are apt to have a considerable degree of mental homogeneity; thus the members of a political meeting are drawn together by common political opinions and sentiments; the audience in a concert room shares a common love of music or a common admiration for the composer, conductor, or great executant; and a still higher degree of homogeneity prevails when a number of persons of the same religious persuasion are gathered together at a great revival meeting. Consider how under such circumstances a very ordinary joke or point made by a political orator provokes a huge delight; how, at a concert, the admiration of the applauding audience swells to a pitch of frantic enthusiasm; how, at the skilfully conducted and successful revival meeting, the fervour of emotion is apt to rise, until it exceeds all normal modes of expression, and men and women give way to loud weeping or even hysterical convulsions.

Such exaltation or intensification of emotion is the most striking result of the formation of a crowd, and is one of the principal sources of the attractiveness of the crowd. By participation in the mental life of a crowd, one's emotions are stirred to a pitch that they seldom or never attain under other conditions. This is for most men an intensely pleasurable experience; they are, as they say, carried out of themselves, they feel themselves caught up in a great wave of emotion, and cease to be aware of their individuality and all its limitations; that isolation of the individual, which oppresses every one of us, though it may not be explicitly formulated in his consciousness, is for the time being abolished. The repeated enjoyment of effects of this kind tends to generate a craving for them, and also a facility in the spread and intensification of emotion in this way; this is probably the principal cause of the greater excitability of urban populations as compared with dwellers in the country, and of the well-known violence and fickleness of the mobs of great cities.

There is one kind of object in the presence of which no man remains indifferent and which evokes in almost all men the same emotion, namely impending danger; hence the sudden appearance of imminent danger may instantaneously convert any concourse of people into a crowd and produce the characteristic and terrible phenomena of a panic. In each man the instinct of fear is intensely excited; he experiences that horrible emotion in full force and is irresistibly impelled to save himself by flight. The terrible driving power of this impulse, excited to its highest pitch under the favouring conditions, suppresses all other impulses and tendencies, all habits of self-restraint, of courtesy and consideration for others; and we see men, whom we might have supposed incapable of cruel or cowardly behaviour, trampling upon women and children, in their wild efforts to escape from the burning theatre, the sinking ship, or other place of danger.

The panic is the crudest and simplest example of collective mental life. Groups of gregarious animals are liable to panic; and the panic of a crowd of human beings seems to be generated by the same simple instinctive reactions as the panic of animals. The essence of the panic is the collective intensification of the instinctive excitement, with its emotion of fear and its impulse to flight. The principle of primitive sympathy[1] seems to afford a full and adequate explanation of such collective intensification of instinctive excitement. The principle is that, in man and in the gregarious animals generally, each instinct, with its characteristic primary emotion and specific impulse, is capable of being excited in one individual by the expressions of the same emotion in another, in virtue of a special congenital adaptation of the instinct on its cognitive or perceptual side. In the crowd, then, the expressions of fear of each individual are perceived by his neighbours; and this perception intensifies the fear directly excited in them by the threatening danger. Each man perceives on every hand the symptoms of fear, the blanched distorted faces, the dilated pupils, the high-pitched trembling voices, and the screams of terror of his fellows; and with each such perception his own impulse and his own emotion rise to a higher pitch of intensity, and their expressions become correspondingly accentuated and more difficult to control. So the expressions of each member of the crowd work upon all other members within sight and hearing of him to intensify their excitement; and the accentuated expressions of the emotion, so intensified,

[1] This principle of primitive sympathy or simple direct induction or contagion of emotion was formulated in Chapter IV of my *Social Psychology*.

react upon him to raise his own excitement to a still higher pitch; until in all individuals the instinct is excited in the highest possible degree.

This principle of direct induction of emotion by way of the primitive sympathetic response enables us to understand the fact that a concourse of people (or animals) may be quickly turned into a panic-stricken crowd by some threatening object which is perceptible by only a few of the individuals present. A few persons near the stage of a theatre see flames dart out among the wings; then, though the flames may be invisible to the rest of the house, the expressions of the startled few induce fear in their neighbours, and the excitement sweeps over the whole concourse like fire blown across the prairie.

The same principle enables us to understand how a few fearless individuals may arrest the spread of a panic. If they experience no fear, or can completely arrest its expressions, and can in any way make themselves prominent, can draw and hold the attention of their fellows to themselves, then these others, instead of perceiving on every hand only the expressions of fear, perceive these few calm and resolute individuals; the process of reciprocal intensification of the excitement is checked and, if the danger is not too imminent and obvious, the panic may die away, leaving men ashamed and astonished at the intensity of their emotion and the violent irrational character of their behaviour.

Other of the cruder primary emotions may spread through a crowd in very similar fashion, though the process is rarely so rapid and intense as in the case of fear.[1]

[1] It was my good fortune to witness the almost instantaneous spread of anger through a crowd of five thousand warlike savages in the heart of Borneo. Representatives of all the tribes of a large district of Sarawak had been brought together by the resident magistrate for the purpose of strengthening friendly relations and cementing peace between the various tribes. All went smoothly, and the chiefs surrounded by their followers

And in every case the principal cause of the intensification of the emotion is the reciprocal action between the members of the crowd, according to the principle of sympathetic induction of emotion in one individual by its expressions in others.

In panic, the dominance of the one emotion and its impulse is so complete as to allow no scope for any of the subtler modes of collective mental operation. But in other cases other conditions co-operate to determine the character of the emotional response of the crowd. Of these the most important are the awareness of the crowd as a whole in the mind of each member of it and his consciousness of his membership in the whole. When a common emotion pervades the crowd, each member becomes more or less distinctly aware of the fact; and this gives him a sense of sharing in a mighty and irresistible power which renders him reckless of consequences and encourages him to give himself up to the prevailing emotion without restraint. Thus, in the case of an audience swept by an emotion of admiration for a brilliant singer, the thunder of applause, which shows each individual that his emotion is shared by all the rest, intensifies his own

---

were gathered together in a large hall, rudely constructed of timber, to make public protestations of friendship. An air of peace and goodwill pervaded the assembly, until a small piece of wood fell from the roof upon the head of one of the leading chiefs, making a slight wound from which the blood trickled. Only the immediate neighbours of this chief observed the accident or could perceive its effect; nevertheless in the space of a few seconds a wave of angry emotion swept over the whole assembly, and a general and bloody fight would have at once commenced, but that the Resident had insisted upon all weapons being left in the boats on the river 200 yards away. The great majority of the crowd rushed headlong to fetch their weapons from their boats, while the few who remained on the ground danced in fury or rushed to and fro gesticulating wildly. Happily the boats were widely scattered along the banks of the river, so that it was possible for the Resident, by means of persuasion, threats, and a show of armed force, to prevent the hostile parties coming together again with their weapons in hand.

emotion, not only by way of sympathetic induction, but also because it frees him from that restraint of emotion which is habitual with most of us in the presence of any critical or adversely disposed spectators, and which the mere thought of such spectators tends to maintain and strengthen. Again, the oratory of a demagogue, if addressed to a large crowd, will raise angry emotion to a pitch of intensity far higher than any it will attain if he is heard by a few persons only; and this is due not only to accentuation of the emotion by sympathetic induction, but also to the fact that, as the symptoms of the emotion begin to be manifested on all sides, each man becomes aware that it pervades the crowd, that the crowd as a whole is swayed by the same emotion and the same impulse as he himself feels, that none remains to criticise the violence of his expressions. To which it must be added that the consciousness of the harmony of one's feelings with those of a mass of one's fellows, and the consequent sense of freedom from all restraint, are highly pleasurable to most men; they find a pleasure in letting themselves go, in being swept away in the torrent of collective emotion. This is one of the secrets of the fascination which draws many thousands of spectators to a football match, and brings together the multitudes of baseball "fans" bubbling over with eager anticipation of an emotional orgy.

The fact that the emotions of crowds are apt to be very violent has long been recognised, and the popular mind, in seeking to account for it, has commonly postulated very special and even supernatural causes. The Negro author of a most interesting book[1] has given the following description of the religious frenzy of a crowd of Christian negroes: "An air of intense excitement possessed the mass of black folk. A suppressed terror hung in the air and seemed to seize us—a pythian madness, a demoniac

[1] *The Souls of Black Folk*, by W. E. B. Du Bois, London, 1905.

possession, that lent terrible reality to song and word. The massive form of the preacher swayed and quivered as the words crowded to his lips. The people moaned and fluttered and then a gaunt brown woman suddenly leaped into the air and shrieked like a lost soul, while round about came wail and groan and outcry, a scene of human passion such as I had never even imagined." The author goes on to say that this frenzy is attributed by the black folk to the direct influence of the Spirit of the Lord, making mad the worshippers with supernatural joy, and that this belief is one of the leading features of their religion. Similar practices, depending upon the tendency of collective emotion to rise to an extreme intensity, have been common to the peoples of many lands in all ages; and similar supernatural explanations have been commonly devised and accepted. I need only remind the reader of the Dionysiac orgies of ancient Greece.

The facts are so striking that for the popular mind they remain unaccountable, and not to be mentioned without some vague reference to magnetism, electricity, hypnotism, or some mysterious contagion; and even modern scientific writers have been led to adopt somewhat extravagant hypotheses to account for them. Thus Dr. Le Bon[1] speaks of "the magnetic influence given out by the crowd" and says that, owing to this influence, "or from some other cause of which we are ignorant, an individual immerged for some length of time in a crowd in action soon finds himself in a special state, which much resembles the state of fascination in which the hypnotised individual finds himself in the hands of the hypnotiser." He goes on to say that in the hypnotised subject the conscious personality disappears and that his actions are the outcome of the unconscious activities of the spinal cord. Now, crowds undoubtedly display great suggestibility, but great sug-

[1] *The Crowd*, p. 11.

gestibility does not necessarily imply hypnosis; and there is no ground for supposing that the members of a crowd are thrown into any such condition, save possibly in very rare instances.

There are however two hypotheses, sometimes invoked for the explanation of the peculiarities of collective mental life, which demand serious consideration and which we may with advantage consider at this point.

One is the hypothesis of telepathy. A considerable amount of respectable evidence has been brought forward in recent years to prove that one mind may directly influence another by some obscure mode of action that does not involve the known organs of expression and of perception; and much of this evidence seems to show that one mind may directly induce in another a state of consciousness similar to its own. If, then, such direct interaction between two minds can take place in an easily appreciable degree in certain instances, it would seem not improbable that a similar direct interaction, producing a lesser, and therefore less easily appreciable, degree of assimilation of the states of consciousness of the minds concerned, may be constantly and normally at work. If this were the case, such telepathic interaction might well play a very important part in collective mental life, and, where a large number of persons is congregated, it might tend to produce that intensification of emotion which is so characteristic of crowds. In fact, if direct telepathic communication of emotion in however slight a degree is possible and normal, and especially if the influence is one that diminishes with distance, it may be expected to produce its most striking results among the members of a crowd; for the emotion of each member might be expected to be intensified by the telepathic influence radiating from every other member. Some slight presumption in favour of such a mode of explanation is afforded by the fact that the popular use of the word contagion in the present connection seems to imply, however vaguely, some such direct communication of emotion. But telepathic communication has not hitherto been indisputably established; and the observations that afford so strong a presumption in its favour indicate that, if and in so far as it occurs, it does so sporadically and only between individuals specially attuned to one another or in some abnormal mental state that renders them specially sensitive to the influence.[1] And, while the acceptance of the principle of sympathetic induction of an emotion, as an instinctive perceptual response to the expressions of that emotion, renders unnecessary any further principle of explanation, the consideration of the conditions of the spread of emotion through crowds affords evidence that this mode of interaction of the individuals is all-important and that telepathic communication, if it occurs, is of secondary importance. For the spreading and the great intensification of emotion seem to depend upon its being given expressions that are perceptible by the senses. So long as its expressions are suppressed, the emotion of an assembly does not become excessive. It is only by eliciting and encouraging the expressions of emotions that the revivalist, the political orator, or the comic man on the music-hall stage, achieves his successes. That the expressions of an emotion are far more effective in this way than the emotion itself is recognised by the practice of the *claqueurs*. When an audience has once been induced to give expression to a common emotion, its members are, as it were, set in tune with one another; each man is aware that he is in harmony with all the rest as regards his feelings and emotions, and, even in the periods during which all expressions are suppressed by the audience, this aware-

[1] In a recent work (*What is Instinct?* by Bingham Newland) the author, who shows an intimate knowledge of the life of wild animals, seems to postulate some such direct telepathic *rapport* between animals of the same species.

ness serves to sustain the mood and to prepare for fresh outbursts. The mere silence of an audience, the absence of coughs, shufflings, and uneasy movements, suffices to make each member aware that all his fellows are attentive and are responding with the appropriate emotion; but it is not until the applause, the indignation, or the laughter, breaks out in free expression that the emotion reaches its highest pitch. And a skilful orator or entertainer, recognising these facts, takes care to afford frequent opportunities for the collective displays of emotion.

We must recognise, then, that, even if telepathic communication be proved to be possible in certain cases, there is not sufficient evidence of its operation in the spread of emotion through crowds, and that the facts are sufficiently explained by another principle of general and indisputable validity, the principle of primitive sympathy.

The second hypothesis to be considered in this connection is that of the "collective consciousness." The conception of a collective consciousness has been reached by a large number of authors along several lines of observation and reasoning and is seriously defended at the present time, more especially by several French and German writers. They maintain that, in some sense and manner, the consciousnesses of individuals are not wholly shut off from one another, but may co-operate in the genesis of, or share in the being of, a more comprehensive consciousness that exists beside and in addition to them. The conception varies according to the route by which it is reached and the use that is made of it; but in all its varieties the conception remains extremely obscure; no one has succeeded in making clear how the relation of the individual consciousness to the collective consciousness is to be conceived. In the writings of many metaphysicians, of whom Hegel is the most prominent, "the Absolute" seems to imply such a collective consciousness, an all-inclusive world-consciousness of which the individual consciousness of each man is somehow but a constituent element or fragmentary manifestation. But it would be unprofitable to attempt any discussion of the conception. We are concerned only with the empirical conception of a collective consciousness based on observation and induction.

Such a conception finds its strongest support in the analogy afforded by a widely current view of the nature and conditions of the psychical individuality of men and animals; the view, namely, that the individual consciousness of any man or animal is the collective consciousness of the cells of which his body, or his nervous system, is composed. We know that the nervous system is made up of cells each of which is a vital unit, capable of living, of achieving its essential vital processes, independent of other cells; and we see free living cells that in many respects are comparable with these and to which we seem compelled, according to the principle of continuity, to attribute some germ of psychical life however rudimentary. What is known of the phylogenetic and ontogenetic development of the multicellular animal seems to justify us in regarding it as essentially an aggregate of such independent vital units, which, being formed by repeated fission from a single cell, adhere together and undergo differentiation and specialisation of functions. If then the parent cell, the germ cell, has a rudimentary psychical life, it is difficult to deny it altogether to the cells formed from it by fission; and it is argued that all these cells continue to enjoy a psychical life and that the consciousness of the individual man or animal is the collective consciousness of some or all of these cells. Now we know that the consciousness of any one of the higher animals has for its physical correlate at any moment processes going on simultaneously in many different parts and elements of the brain. It is argued, then, that we must suppose each cell of the brain to enjoy, whenever it is active, its own psychical life, and at the same time to contribute something towards the unitary "collective consciousness" of the whole organism, which thus exists beside, but not independent of, these rudimentary consciousnesses of the cells. If the view be accepted, it affords a close analogy with the supposed "collective consciousness" of a group of men or a society.

This conception of the collective nature of the consciousness of complex organisms finds strong support in two classes of facts. First, it finds support in the fact that, if individuals of many of the animal species of an intermediate grade of complexity, such as some of the worms and some of the radiate animals, be cut into two or more parts, each part may continue to live and may become a complete organism by reconstitution of the lost parts. Since, then, we can hardly deny some integrated psychical life to such organism, some rudimentary consciousness, we seem compelled to believe that this consciousness may be divided into two or more consciousnesses, each of them being associated with the vital activities of one of the parts into which the organism is divided by the knife. Division of the organism into two parts is also the normal mode of reproduction in the animal world. Even the coming into existence of every human being seems to be bound up with the separation of a cell from the parent organism; and his existence as a separate psychical individual seems to result from the same process of physical division. And if one cell, when thus separated from the parent organism, can thus prove its possession of a psychical life by developing into a fully conscious organism, it is difficult to deny that all other cells have also their own psychical lives, even though they may be incapable of making it manifest to us by growing up into complex organisms when separated.

The second class of facts that seem to justify this conception of the consciousness of complex organisms are facts which have been studied and discussed widely in recent years under the head of mental dissociation or disintegration of personalities. Such disintegration seems to occur spontaneously as the essential feature of severe hysteria, and to be producible artificially and temporarily in some subjects, when they are thrown into deep hypnosis. In certain of these cases the behaviour of the human being seems to imply that it is the expression of two separate psychical individuals, formed by the splitting of the stream of consciousness and of mental activity of the individual into two streams. The two streams may be of co-ordinate complexity; but more frequently one of them seems to be a mere trickle diverted from the main stream of personal consciousness. Since it is, from the nature of the case, always impossible to obtain any direct and certain proof that any behaviour other than one's own is the expression of conscious mental processes, it is not possible to prove that such division or disintegration of the personal consciousness actually takes place. But the facts appear to many of the psychologists who have studied them most carefully[1] to demand this interpretation; and this psychical disintegration seems to be accompanied by a functional dissociation of the nervous system into two or more systems each of which functions independently of the others, —that is to say, a division of the nervous system comparable with the division of the nervous system of the worm by the stroke of the knife which seems to split the psychical individual into two.

The facts of both these orders would appear, then, to indicate that the physical organisation of the cells of a

[1] See *The Dissociation of a Personality*, by Dr. Morton Prince; *Double Personality*, by A. Binet; *The Psychology of Suggestion*, by Boris Sidis; *L'automatism psychologique*, by Pierre Janet; and the descriptions and discussions of William James in his *Principles of Psychology*.

complex organism is accompanied by an organisation of their psychical lives to form a "collective consciousness," which in the human being becomes a personal self-consciousness; and they would seem to show that the unity of personal consciousness has for its main condition the functional continuity of the protoplasm of the cells of the nervous system.

Even before the facts of disintegration of personalities were known, several authors, notably von Hartmann[1] and G. T. Fechner,[2] did not hesitate to make this last assumption; and to assert that, if the brain of a man could be divided by a knife into two parts each of which continued to function, his consciousness would thus be divided into two consciousnesses; and conversely, that, if a functional bridge of nervous matter could be established between the brains of two men, their consciousnesses would fuse to a single consciousness. The discovery of these facts has greatly strengthened the case for this view; and it has been accepted by so sound a psychologist and sober a philosopher as Fouillée.[3]

It may be claimed that the consideration of the nature and behaviour of animal societies points to a similar conclusion, and supplements in an important manner the argument founded on the divisibility of individual organisms. Such a line of reasoning has been most thoroughly pursued by Espinas in his very interesting book on animal societies.[4] He begins by considering the lower polycellular forms of animal life. Among them, especially among the hydrozoa or polypes, we find compound or colonial animals; such an animal is a single living mass of which all the parts are in substantial and vital connection with one another, but is yet made up of a number of parts each of which is morphologically a complete or almost

[1] *Philosophy of the Unconscious.*
[2] *Die Psychophysik.*
[3] *Psychologie des idées forces.*
[4] *Les Sociétés animales*, Paris, 1877.

complete creature; and these parts, though specialised for the performance of certain functions subserving the economy of the whole animal or coherent group of animals, are yet capable, if separated from the mass (as they sometimes are by a natural process), of continuing to live, of growing, and of multiplying. There are found among such creatures very various degrees of specialisation of parts and of interdependence of parts; and in those cases in which the specialisation and interdependence of parts is great, the whole compound animal exhibits in its reactions so high a degree of integration that we seem justified in supposing that a common or "collective consciousness" is the psychical correlate of these integrated actions of the separable parts. Why then, it is asked, should this "collective consciousness" cease to be, when the substantial continuity of the parts is interrupted?

Espinas then goes on to describe animal societies of many types, and shows how, as we follow up the evolutionary scale, association and intimate interdependence and co-operation of their members tend to replace more and more completely the individualistic antagonism and unmitigated competition of the lowest free-living organisms. He considers first the type of animal society which is essentially a family, a society of individuals all of which are derived from the same parent by fission or by budding. He argues that each such society of blood-relatives is a harmonious whole only because it enjoys a "collective consciousness" over and above the consciousnesses of its constituent members; that, for example, a swarm of bees, which exhibits so great a uniformity of feeling and action and of which all the members come from the body of one parent, is in reality the material basis of a "collective consciousness," which presides over and is expressed by their collective actions; that the ants of one household have such a collective consciousness, that they "are, in truth, a single thought in action, like the various cellules and fibres of the brain of a mammal." For, as he maintains, "the consciousness of animals is not an absolute, indivisible thing. It is on the contrary a reality capable of being divided and diffused . . . thought in general and the impulses illuminated by it, are, like the forces of nature, susceptible of diffusion, of transmission, of being shared, and can like these lie dormant where they are thinly diffused, or become vivid and intensified by concentration. The beings that have these attributes are no doubt monads; but these monads are open to and communicate with one another."

Espinas extends the view to other animal societies of which the members are not all derived from one parent, including human societies; and concludes that, except in the case of the Infusoria at the bottom of the scale and of the highly organised societies at the top of it, every individual consciousness is a part of a superior more comprehensive consciousness of an individual of a higher order. He illustrates at length the fact with the consideration and explanation of which this chapter is concerned, the fact namely that, in all social groups, emotions and impulses are communicated and intensified from one individual to another; and he asks—"If the essential elements of consciousness add themselves together and accumulate from one consciousness to another, how should the consciousness itself of the whole not be participated in by each?" He argues that to be real is not to be known to some other consciousness, but is to exist for oneself, to be conscious of oneself; that, in this sense, the "collective consciousness" of a society is the most real of all things; that every society is therefore a living individual; and that, if we deny self-conscious individuality to a society, we must deny it equally to the mass of cells that make up an animal body; that, in short, we can find unity and individuality nowhere.

This doctrine of the "collective consciousness" of

societies may seem bizarre to those to whom it is altogether novel; but it is one that cannot be lightly put aside; it demands serious consideration from any one who seeks the general principles of Collective Psychology. We have no certain knowledge from which its impossibility can be deduced; and the new light thrown upon individuality by modern studies in psycho-pathology shows us that the indivisibility and strictly bounded unity of the individual human soul is a postulate that we must not continue to accept without critical examination. Nor is the conception one that figures only in the writings of philosophers and therefore to be regarded with contemptuous indulgence by men of affairs as but one of the strange, harmless foibles of such persons. It has a certain vogue in more popular writings; thus Renan wrote—"It has been remarked that in face of a peril a nation or a city shows, like a living creature, a divination of the common danger, a secret sentiment of its own being and the need of its conservation. Such is the obscure impulsion which provokes from time to time the displacement of a whole people or the emigration of masses, the crusades, the religious, political, or social revolutions." Phrases such as the soul of a people, the genius of a people, have long been current, and in almost every newspaper one may find important events and tendencies ascribed to the instinct of a people. It is probable that these phrases are written in many instances without any explicit intention to imply a "collective national consciousness," but merely as well-sounding words that cloak our ignorance and give a vague appearance of understanding. Nevertheless, from its application to the life of nations, the doctrine of a collective consciousness mainly derives its importance. It is seriously used by a number of vigorous contemporary writers, of whom Schaeffle[1] is perhaps the most notable,

[1] *Bau und Leben des Socialen Körpers.*

to carry to its extreme the doctrine of Comte and Spencer that Society is an organism. Spencer specifically refused to complete his analogy between society and an animal organism by the acceptance of the hypothesis of a collective consciousness; and he insisted strongly on the importance, for legislation and social effort of every kind, of holding fast to the consciousness of individual men as the final court of appeal, by reference to which the value of every institution and every form of social activity must be judged; the importance of regarding the welfare and happiness of individual men as the supreme end, in relation to which the welfare of the State is but a means. But those who, like Schaeffle, complete the analogy by acceptance of this hypothesis, regard a nation as an organism in the fullest sense of the word, as an organism that has its own pleasure and pain and its own conscious ends and purposes and strivings; as in fact a great individual which is conscious and may be more or less perfectly self-conscious, conscious of itself, its past, its future, its purposes, its joys and its sorrows. And they do not scruple to draw the logical conclusion that the welfare of the individual should be completely subjected to that of the State; just as the welfare of an organ or cell of the human body is rightly held to be of infinitesimal value in comparison with that of the whole individual and to derive its importance only from its share in the constitution of the whole. This conception of the "collective consciousness" has thus been used as one of the supports of "Prussianism" and has played its part in bringing about the Great War with all its immense mass of individual anguish.

We must, then, examine the arguments upon which the doctrine is based, and ask—Do they suffice to render it probable, or to compel our acceptance of it, and to justify the complete subjection of the individual to the State?

We have seen that a strong case is made out for the view

that the consciousness of a complex organism is the "collective consciousness" of all its cells, or of the cells of its nervous system; and it must be admitted that, if this view could be definitely established, it would go far to justify the doctrine of the collective consciousness of societies. Yet the view is by no means established; there are great difficulties in the way of its acceptance. There is the difficulty which meets a doctrine of "collective consciousness" in all its forms from that of Haeckel to that of Hegel, —the difficulty that the consciousness of the units is used twice over, once as the individual consciousness, once as an element entering into the collective consciousness; and no one has been able to suggest how this difficulty can be surmounted. It has been argued also, most forcibly perhaps by Lotze,[1] that what we know of the structure and functions of the brain compels us to adopt a very different interpretation of the facts. It is said that, since we cannot find any evidence of a unitary brain-process that might be regarded as the immediate physical correlate of the unitary stream of consciousness of the individual, but find rather that the physical correlate of the individual's consciousness at any moment is a number of discrete processes taking place simultaneously in anatomical elements widely scattered in different parts of the brain, we are compelled to assume that each of these acts upon some unitary substance, some immaterial entity (which may be called the soul) producing a partial affection of its state. According to this view, then, the consciousness of any moment is the unitary resultant of all these influences simultaneously exerted on the soul, the unitary reaction of the soul upon these many influences.[2]

[1] *Medicinische Psychologie.*

[2] I have argued that the great increase of knowledge of the functions and structure of the nervous system attained by recent research does but provide for the argument a surer basis of empirical data; and I have contended that

But, even if we could accept the view that the consciousness of the complex organism is the "collective consciousness" of its cells, the analogy between an organism and a society, which constitutes the argument for the "collective consciousness" of a society, would remain defective in one very important respect. If we accept that view, we must believe that the essential condition of the fusion of the consciousnesses of the cells is their spatial continuity, no matter how utterly unintelligible this condition may seem; for the apparent disruption of consciousness on the solution of material continuity between the cells is the principal ground on which this view is founded. Now, no such continuity of substance exists between the members of any human group or society, and its absence constitutes a fatal flaw in the analogical argument.

If we pass by these serious difficulties, others arise as soon as we inquire what kinds of human groups have such "collective consciousness." Does the simple fortuitously gathered crowd possess it? Or is it confined to highly organised groups such as the leading modern nations? If every psychological crowd possesses it and owes its peculiarities of behaviour to it, does it come into being at the moment the individuals have their attention attracted to a common object and begin to be stirred by a common emotion? And does it cease to be as soon as the crowd is resolved into its elements? Or, if it is confined to nations or other highly organised groups, at what stage of their development does it come into being, and what are the limits of the groups of which it is the "collective conscious-

---

some at least of the cases of disintegration of personality are more easily reconcilable with this view than with the contrary doctrine which regards the individual consciousness as the collective consciousness of the brain-cells. See my *Body and Mind*, a book I found myself compelled to write in order to arrive at a reasoned judgment on this difficult problem, which obtrudes itself at the outset of the study of group life.

ness?" Do the Poles share in the "collective consciousness" of the German nation, or the Bavarians in that of Prussia? Or do the Irish or the Welsh contribute their share to that of the English nation?

Coming now to close quarters with the doctrine, we may ask those who, like Schaeffle and Espinas, regard the "collective consciousness" as a bond which unites the members of a society and makes of them one living individual—Is this "collective consciousness" merely epiphenomenal in character? Or are we to regard it as reacting upon the consciousnesses or minds of the individuals of the group, and, through such reaction, playing a part in determining the behaviour of the group, or rather of the individuals of which the group is composed? For the actions of the group are merely the sum of the actions of its individuals. If the former alternative be adopted, then we may confidently say that the existence of a "collective consciousness" must from the nature of the case remain a mere speculation, incapable of verification; and that, if it does exist, since it cannot make any difference, cannot in any way affect human life and conduct, it is for us unreal, no matter how real it may be for itself, as Espinas maintains; and we certainly are not called upon to have any regard for it or its happiness, nor can we invoke its aid in attempting to explain the course of history and the phenomena of social life. If, on the other hand, the "collective consciousness" of groups and societies and peoples reacts upon individual minds and so plays a part in shaping the conduct of men and societies, then the conception is a hypothesis which can only be justified by showing that it affords explanations of social phenomena which in its absence remain inexplicable. If it were found that social aggregates of any kind really do exhibit, as has often been maintained, great mass-movements, emigrations, religious or political uprisings, and so forth, for which no adequate explanations can be found in the mental processes of individuals and the mental interactions of individuals by the ordinary means of expression and perception, a resort to some such hypothesis would be permissible; but it is an offence against the principles of scientific method to invoke its aid, before we have exhausted the possibilities of explanation offered by well-known existents and forces. That certainly has not yet been done, and the upholders of the doctrine have hardly made any attempt to justify it in this, the only possible, manner in which it could be justified. The only evidence of this sort adduced by Espinas is the rapid spread of a common emotion and impulse throughout the members of animal and human groups; and of such phenomenon we have already found a sufficient explanation in those special adaptations of the instincts of all gregarious creatures which are unmistakably implied by the way in which the expression of an emotion directly evokes a display of the same emotion in any onlooking member of the species.

We may, then, set aside the conception of a "collective consciousness" as a hypothesis to be held in reserve until the study of group life reveal phenomena that cannot be explained without its aid. For it may be confidently asserted that up to the present time no such evidence of a "collective consciousness" has been brought forward, and that there is no possibility of any such evidence being obtained before the principles of social psychology have been applied far more thoroughly than has yet been done to the explanation of the course of history. In adopting a so far unsympathetic attitude towards this doctrine, we ought to admit that, if there be any truth in it, the "collective consciousness" of even the most highly organised society may be still in a rudimentary stage, and that it may continue to gain in effectiveness and organisation with the further evolution of the society in question.

After this digression we may return to the consideration of the emotional characteristics of simple crowds. We have to notice not only that the emotions of crowds are apt to be excessively strong, but also that certain types of emotion are more apt than others to spread through a crowd, namely the coarser, simpler emotions and those which do not imply the existence of developed and refined sentiments. For many of the individuals of most crowds will be incapable of the more subtle complex emotions and will be devoid of the more refined sentiments; while such sentiments as the individuals possess will be in the main more diverse in proportion to their refinement and special character; hence the chances of any crowd being homogeneous as regards these emotions and sentiments is small. Whereas the primary emotions and the coarser sentiments may be common to all the members of a crowd; any crowd is likely to be homogeneous in respect to them.

On the other hand, a crowd is more apt to be swayed by the more generous of the coarser emotions, impulses, and sentiments than by those of a meaner, universally reprobated kind. For each member of the crowd acts in full publicity; and his knowledge of, and regard for, public opinion will to some extent incline him to suppress the manifestation of feelings which he might indulge in private but would be ashamed of in public. Hence a crowd is more readily carried away by admiration for a noble deed, or by moral indignation against an act of cruelty, than by self-pity or jealousy or envy or a meanly vengeful emotion.

At the same time, a crowd is apt to express feelings which imply less consideration and regard for others than the individual, representing the average morality and refinement of its members, would display when not under the influence of the crowd. Thus men, when members of a crowd, will witness with enjoyment scenes of brutality and suffering which, under other circumstances, they would turn away from, or would seek to terminate. To see a man thrown heavily to the ground is not pleasing to most individuals; yet the spectacle provokes roars of delight from the crowd at a football match. How many of the spectators, who, as members of a crowd, hugely enjoy looking on at a prize-fight or a bullfight, would shrink from witnessing it as isolated individuals! How many boys will join with a crowd of others in cruelly teasing another boy, an animal, an old woman, or a drunken man, who individually are incapable of such "thoughtless" conduct! It may be doubted whether even the depraved population of Imperial Rome could have individually witnessed without aversion the destruction of Christians in the Coliseum.

This character of crowds seems to be due to two peculiarities of the collective mental state. In the first place, the individual, in becoming one of a crowd, loses in some degree his self-consciousness, his awareness of himself as a distinct personality, and with it goes also something of his consciousness of his specifically personal relations; he becomes to a certain extent depersonalised. In the second place, and intimately connected with this last change, is a diminution of the sense of personal responsibility: the individual feels himself enveloped and overshadowed and carried away by forces which he is powerless to control; he therefore does not feel called upon to maintain the attitude of self-criticism and self-restraint which under ordinary circumstances are habitual to him, and his more refined ideals of behaviour fail to assert themselves against the overwhelming forces that envelop him.

The Intellectual Processes of Simple Crowds

No fact has been more strongly insisted upon by writers on the psychology of crowds than the low degree of intelligence implied by their collective actions. Not only mobs or simple crowds, but such bodies as juries, committees, corporations of all sorts, which are partially organised groups, are notoriously liable to pass judgments, to form decisions, to enact rules or laws, so obviously erroneous, unwise, or defective that any one, even the least intelligent member of the group concerned, might have been expected to produce a better result.

The principal ground of the low order of intelligence displayed by simple crowds is that the ideas and reasonings which can be collectively understood and accepted must be such as can be appreciated by the lower order of minds among the crowd. These least intelligent minds bring down the intelligence of the whole to their own level. This is true in some degree even of crowds composed of highly educated persons; for, as in the case of the emotions and sentiments, the higher faculties are always more or less specialised and differentiated in various ways through differences of nurture and training; whereas the simpler intellectual faculties and tendencies are common to all men.

A second condition, which co-operates with the foregoing to keep the intellectual processes of crowds at a low level, is the increased suggestibility of its members. Here is one of the most striking facts of collective mental life. A crowd impresses each of its members with a sense of its power, its unknown capacities, its unlimited and mysterious possibilities; and these, as I have shown in Chapter III of my *Social Psychology*, are the attributes that excite in us the instinct of subjection and so throw us into the receptive suggestible attitude towards the object that displays them. Mere numbers are capable of exerting this effect upon most of us; but the effect of numbers is greatly increased if all display a common emotion and speak with one voice; the crowd has then, if we are in its presence, a well-nigh irresistible prestige. Hence even the highly intelligent and self-reliant member of a crowd is apt to find his critical reserve broken down; and, when an orator makes some proposition which the mass of the crowd applauds but which each more intelligent member would as an individual reject with scorn, it is apt to be uncritically accepted by all alike; because it comes to each, not as the proposition of the orator alone, but as a proposition which voices the mind of the crowd, which comes from the mass of men he sees around him and so comes with the power of a mass-suggestion.

A further ground of the suggestibility of the crowd is that prevalence of emotional excitement which was discussed in the foregoing pages. It is well recognised that almost any emotional excitement increases the suggestibility of the individual, though the explanation of the fact remains obscure. I have suggested that the explanation is to be found in the principle of the vicarious usage of nervous energy, the principle that nervous energy, liberated in any one part of the nervous system, may overflow the channels of the system in which it is liberated and re-enforce processes initiated in other systems. If this be true, we can see how any condition of excitement will favour suggestibility; for it will re-enforce whatever idea or impulse may have been awakened and made dominant by "suggestion." The principle requires perhaps the following limitation. Emotion which is finding outlet in well-directed action is probably unfavourable to all such "suggestions" as are not congruent with its tendencies. It is vague emotion, or such as finds no appropriate expression in action, that favours suggestibility. The most striking illustrations of the greatly increased suggestibility of crowds are afforded by well-authenticated instances of collective hallucination, instances which, so long as we fail to take into account the abnormal suggestibility of the members of crowds, seem utterly mysterious, incredible, and super-normal.

Again, the capacity of crowds to arrive at correct conclusions by any process of reasoning is apt to be diminished in another way by the exaltation of emotion to which, as we have seen, they are peculiarly liable. It is a familiar fact that correct observation and reasoning are hampered by emotion; for all ideas congruent with the prevailing emotion come far more readily to consciousness and persist more stably than ideas incongruent with it, and conclusions congruent with the prevailing emotion and desire are accepted readily and uncritically; whereas those opposed to them can hardly find acceptance in the minds of most men, no matter how simple and convincing be the reasoning that leads to them.

The diminution or abolition of the sense of personal responsibility, which results from membership in a crowd and which, as we have seen, favours the display of its emotions, tends also to lower the level of its intellectual processes. Wherever men have to come to a collective decision or to undertake collective action of any sort, this effect plays an important part. The weight of responsibility that would be felt by any one man, deciding or acting alone, is apt to be divided among all the members of the group; so that for each man it is diminished in proportion to the number of persons taking part in the affair. Hence the attention and care devoted by each man to the task of deliberation, observation, or execution, are less keen and continuously sustained, and a judgment or decision is more lightly and easily arrived at, grounds which the individual, deliberating alone, would reject or weigh again and again serving to determine an immediate judgment. The principle is well recognised in practical life. We do not set ten men to keep the look-out on ship-board, but only one; though the safety of the ship and of all that it carries depends upon his unremitting alertness. We see the principle recognised in the institution of the jury. But for the weakening of the individual sense of responsibility, juries would seldom be found capable of finding a prisoner guilty of murder and so condemning him to death; while, by the restriction of the jury to a comparatively small number, the worst features of collective mental life are avoided.

We see the working of the principle not only in simple crowds, but also in groups of very considerable degrees of organisation. We see it in the way in which many a man, who would shrink from the responsibility of directing a great and complicated commercial undertaking, will cheerfully join a board of directors each of whom is perhaps no better qualified than himself to conduct the business of the concern. We may recognise its effects also in the cheerful levity, not to say hilarity, that frequently pervades our House of Commons; for most of its well-meaning members would be utterly crushed under the weight of their legislative responsibility, were it not divided in small fractions among them.

But the low sense of responsibility of the crowd is not due to the division of responsibility alone. In the case of the simple crowd, it is due also in large part to the fact that such a crowd has but a very low grade of self-consciousness and no self-regarding sentiment; that is to say, the members of the crowd have but a dim consciousness of the crowd as a whole, but very little knowledge of its tendencies and capacities, and no sentiment of love, respect, or regard of any kind for it and its reputation in the eyes of men. Hence, since the responsibility falls on the whole crowd, and any loss or gain of reputation affects the crowd and hardly at all the individuals who are merged in it, they are not stimulated to exert care and self-restraint and critical deliberation in forming their judgments, in arriving at decisions, or in executing any task collectively undertaken. The results of these two conditions of collective mental life are well summed up in the popular dictum that a corporation has no conscience.

Since all these factors co-operate to keep the intellectual activity of the simple crowd on a low level, it follows that very simple intellectual processes must be relied on by the orator who would sway a crowd; he must rely on abuse and ridicule of opponents, or unmeasured praise of friends; on flattery; on the *argumentum ad hominem;* on induction by simple enumeration of a few striking instances; on obvious and superficial analogies; on the evocation of vivid representative imagery rather than of abstract ideas; and, above all, on confident assertion and re-iteration, and on a display of the coarser emotions.

Since the individuals comprised in a crowd are apt to be influenced in all these ways by the mass of their fellows, it follows that the mental processes, the thought and feelings and actions, of each one will be as a rule very different from what they would be if he faced a similar situation as an isolated individual; the mental processes of each one are profoundly modified by his mental interactions with all the other members of the crowd. Therefore the collective actions of a crowd are not simply the resultants of all the tendencies to thought and action of the individuals, as such, but may be very different from any such resultant. And they are not merely the expression of the individual tendencies of the average member, nor yet of the mass of least intelligent and refined members; they may be, and often are, such as no one of the members acting alone would ever display or attempt.

It must be added that all the peculiarities of collective mental process mentioned above express themselves very readily in the actions of simple crowds, because such a crowd is incapable of resolution and volition in the true sense of the words. I have shown[1] that individual resolution and volition are only rendered possible by the possession of a well-developed self-consciousness and self-regarding sentiment. But a simple crowd has at the most only a rudimentary self-consciousness and has no self-regarding sentiment. Hence its actions are the direct issue of the various impulses that are collectively evoked; and, though it may be collectively conscious of the end towards which it is impelled, and though all the individuals may desire to effect or realise this end, and to that extent may be said to be capable of purpose; yet such an impulse or desire cannot be steadied, strengthened, renewed, or supported and maintained, in opposition to any other impulse that may come into play, by an impulse springing from the self-regarding sentiment in the way which constitutes resolution and volition. Just so far as the self-regarding sentiment of individuals comes into play and they exert their individual volitions, they cease to act as members of a crowd. The actions of the simple crowd are thus not the outcome of a general will, nor are they the resultant of the wills of all its members; they are simply not volitional in the true sense, but rather impulsive. They are comparable with the actions of an animal rather than with those of a man. It is the lack of the conditions necessary to collective resolution and volition that renders a crowd so fickle and inconsistent; so capable of passing from one extreme of action to another, of hurrying to death the man whom it glorified at an earlier moment, or of turning from savage butchery to tender and tearful solicitude. Such incapacity of the crowd for resolution and volition, together with the increased suggestibility of its members, accounts for the fact that a crowd may be easily induced to follow as a leader any one who, by means of the elementary reasoning processes suited to its intellectual capacity, can suc-

[1] *Social Psychology*, Chapter IX.

ceed in suggesting to it the desirability of any course of action.

We may sum up the psychological character of the unorganised or simple crowd by saying that it is excessively emotional, impulsive, violent, fickle, inconsistent, irresolute and extreme in action, displaying only the coarser emotions and the less refined sentiments; extremely suggestible, careless in deliberation, hasty in judgment, incapable of any but the simpler and imperfect forms of reasoning; easily swayed and led, lacking in self-consciousness, devoid of self-respect and of sense of responsibility, and apt to be carried away by the consciousness of its own force, so that it tends to produce all the manifestations we have learnt to expect of any irresponsible and absolute power. Hence its behaviour is like that of an unruly child or an untutored passionate savage in a strange situation, rather than like that of its average member; and in the worst cases it is like that of a wild beast, rather than like that of human beings.

All these characteristics of the crowd were exemplified on a great scale in Paris at the time of the great Revolution, when masses of men that were little more than unorganised crowds escaped from all control and exerted supreme power; and writers on the topic have drawn many striking illustrations from the history of the days of the Terror.[1] The understanding of these more elementary facts and principles of group psychology will prevent us falling into such an error as was committed by our greatest political philosopher, Edmund Burke, when he condemned the French people in the most violent terms on account of the terrible events of the Revolution; for he attributed to the inhabitants of France in general, as individuals, the capaci-

[1] See especially A. Stoll's *Suggestion und Hypnotismus in der Völkerpsychologie,* where the events of the French Revolution have been treated in some detail from this point of view.

ties for violence and brutality and the gross defects of intelligence and self-restraint that were displayed by the Parisian crowds of the time; whereas the study of collective psychology has led us to see that the actions of a crowd afford no measure of the moral and intellectual status of the individuals of which it is composed. So, when we hear of minor outrages committed by a crowd of undergraduates or suffragettes, a knowledge of group psychology will save us from the error of attributing to the individuals concerned the low grade of intelligence and decency that might seem to be implied by the deeds performed by them collectively. The same understanding will also resolve for us some seeming paradoxes; for example, the paradox that, while in the year 1906 the newspapers contained many reports of almost incredible brutalities committed by the peasants in many different parts of Russia, an able correspondent, who was studying the peasants at that very time, ascribed to them, as the most striking quality of their characters, an exceptional humaneness and kindliness.[1]

It will be maintained on a later page that we may properly speak not only of a collective will, but also of the collective mind of an organised group, for example, of the mind and will of a nation. We must, then, ask at this stage—Can we properly speak of the collective mind of an unorganised crowd? The question is merely one as to the proper use of words and therefore not of the first importance. If we had found reason to accept the hypothesis of a "collective consciousness" of a group, and to believe that the peculiarities of behaviour of a crowd are due to a "collective consciousness," then we should certainly have to admit the propriety of regarding the crowd as having a collective mind. But we have provisionally rejected that hypothesis, and have maintained that the

[1] Hon. Maurice Baring in an article in the *Morning Post* of April 21, 1906.

5

only consciousness of a crowd or other group is the consciousnesses of its constituent individuals. In the absence of any "collective consciousness" we may still speak of collective minds; for we have defined a mind as an organised system of interacting mental or psychical forces. This definition, while allowing us to speak of the collective mind of such a group as a well-developed nation, hardly allows us to attribute such a mind to a simple crowd: for the interplay of its mental forces is not determined by the existence of an organised system of relations between the elements in which the forces are generated; and such determination is an essential feature of whatever can be called a mind.

CHAPTER III

The Highly Organised Group

THE peculiarities of simple crowds tend to appear in all group life; but they are modified in proportion as the group is removed in character from a simple crowd, a fortuitous congregation of men of more or less similar tendencies and sentiments. Many crowds are not fortuitous gatherings, but are brought together by the common interest of their members in some object or topic. These may differ from the simple fortuitous crowd only in being more homogeneous as regards the sentiments and interests of their members; their greater homogeneity does not in itself raise them above the mental level of the fortuitous crowd; it merely intensifies the peculiarities of group life, especially as regards the intensity of the collective emotion.

There is, however, one condition that may raise the behaviour of a temporary and unorganised crowd to a higher plane, namely the presence of a clearly defined common purpose in the minds of all its members. Such a crowd, for example a crowd of white men in one of the Southern States of North America setting out to lynch a negro who is supposed to have committed some flagrant crime, will display most of the characteristics of the common crowd, the violence and brutality of emotion and impulse, the lack of restraint, the diminished sense of responsibility, the increased suggestibility and incapacity for arriving at correct conclusions by deliberation and the weighing of evidence. But it will not exhibit the fickleness of a common crowd, the easy yielding to distracting impressions and to suggestions that are opposed to the common purpose. Such a crowd may seize and execute its victim with inflexible determination, perhaps with a brutality and a ruthless disregard of all deterrent considerations of which no one of its members would be individually capable; and may then at once break up, each man returning quietly and seriously to his home, in a way which has often been described by witnesses astonished at the contrast between the behaviour of the crowd and that of the individuals into which it suddenly resolves itself.

The behaviour of a crowd of this kind raises the problem of the general or collective will. It was said in the foregoing chapter that the actions of a common crowd cannot properly be regarded as volitional, because they are the immediate outcome of the primary impulses. Yet the actions of a crowd of the kind we are now considering are the issue of true resolutions formed by each member of the crowd, and are, therefore, truly volitional. Nevertheless, they are the expression not of a general or collective will, but merely of the wills of all the individuals; and, even if there arise differences between the members and a conflict of wills as to the mode of achieving the common end, and if the issue be determined simply by the stronger party overbearing the weaker and securing their co-operation, that still does not constitute the expression of a general will. For a collective or general will only exists where some idea of the whole group and some sentiment for it as such exists in the minds of the persons composing it. But we may with advantage examine the nature of collective volition on a later page, in relation to the life of a highly organised group, such as an army.

There are five conditions of principal importance in raising collective mental life to a higher level than the unorganised crowd can reach, no matter how homogeneous the crowd may be in ideas and sentiments nor how convergent the desires and volitions of its members. These are the principal conditions which favour and render possible the formation of a group mind, in addition to those more fundamental conditions of collective life which we have noted in the foregoing chapter.

The first of these conditions, which is the basis of all the rest, is some degree of continuity of existence of the group. The continuity may be predominantly material or formal; that is to say, it may consist either in the persistence of the same individuals as an inter-communicating group, or in the persistence of the system of generally recognised positions each of which is occupied by a succession of individuals. Many permanent groups exhibit both forms of continuity in a certain degree; for, the material continuity of a group being given, some degree of formal continuity will commonly be established within it. The most highly organised groups, such as well-developed nations, exhibit both forms in the highest degree.

A second very important condition, essential to any highly developed form of collective life, is that in the minds of the mass of the members of the group there shall be formed some adequate idea of the group, of its nature, composition, functions, and capacities, and of the relations of the individuals to the group. The diffusion of this idea among the members of the group, which constitutes the self-consciousness of the group mind, would be of little effect or importance, if it were not that, as with the idea of the individual self, a sentiment of some kind almost inevitably becomes organised about this idea and is the main condition of its growth in richness of meaning; a sentiment for the group which becomes the source of emotions and of impulses to action having for their objects the group and its relations to other groups.

A third condition very favourable to the development of the collective mind of a group, though not perhaps absolutely essential, is the interaction (especially in the form of conflict and rivalry) of the group with other similar groups animated by different ideals and purposes, and swayed by different traditions and customs. The importance of such interaction of groups lies chiefly in the fact that it greatly promotes the self-knowledge and self-sentiment of each group.

Fourthly, the existence of a body of traditions and customs and habits in the minds of the members of the group determining their relations to one another and to the group as a whole.

Lastly, organisation of the group, consisting in the differentiation and specialisation of the functions of its constituents—the individuals and classes or groups of individuals within the group. This organisation may rest wholly or in part upon the conditions of the fourth class, traditions, customs, and habits. But it may be in part imposed on the group and maintained by the authority of some external power.

The capacity for collective life of an organised group whose organisation is imposed upon it and wholly maintained by an external authority is but little superior to that of a simple crowd. Such a group will differ from the simple crowd chiefly in exhibiting greater control of its impulses and a greater continuity of direction of its activities; but these qualities are due to the external compelling power and are not truly the expression of its collective mental life. An army of slaves or, in a less complete degree, an army of mercenaries is the type of this kind of organised group; and a people ruled by a strong despot relying on a mercenary or foreign army approximates to it. The first aim of the power that would maintain such an organisation must always be to prevent and suppress collective life, by forbidding gatherings and public discussions, by rendering communications between the parts difficult, and by enforcing a rigid discipline. For such an organisation is essentially unstable.

We may illustrate the influence of these five conditions by considering how in a group of relatively simple kind, in which they are all present, they favour collective life and raise it to a higher level of efficiency. Such a group is a patriot army fighting in a cause that elicits the enthusiasm of its members; such were the armies of Japan in the late Russo-Japanese war; they exhibited in a high degree and in relative simplicity the operation of all the conditions we have enumerated.

Such an army exhibits the exaltation of emotion common to all psychological crowds. This intensification of emotion enables men to face danger and certain death with enthusiasm, and on other occasions may, even in the armies of undoubtedly courageous and warlike nations, result in panic and a rout. But in all other respects the characteristics of the simple crowd are profoundly modified. The formal continuity of the existence of the army and of its several units secures for it, even though its personnel be changed at a rapid rate, a past and therefore a tradition, a self-consciousness and a self-regarding sentiment, a pride in its past and a tradition of high conduct and achievement; for past failures are discreetly forgotten and only its past successes and glories are kept in memory. The traditional group consciousness and sentiment are fostered by every wise commander, both in the army as a whole and in each separate department and regiment. Is not the superiority in battle of such bodies as the famous Tenth Legion due as much to such self-conscious tradition and sentiment as to the presence of veterans in its ranks? And is not the same true of such regiments as the Black Watch, the Gordons, the Grenadier Guards, and the other famous regiments of the British army?

The third of the conditions mentioned above is also very obviously present in the case of an army in the field—namely, interaction with a similar group having different purposes, traditions, and sentiments. And in this case the interaction, being of the nature of direct competition and conflict, is of the kind most favourable to the development of the collective mind. It accentuates the self-consciousness of the whole; that is to say, it defines more clearly in the mind of each individual the whole of which he is a part, his position in, his organic connection with, and his dependence upon, the whole; with each succeeding stage of the conflict he conceives the whole more clearly, obtains a fuller knowledge of the capacities and weaknesses of the whole and its parts. Each soldier learns, too, something of the character of the opposing army; and, in the light of this knowledge, his conception of his own army becomes better defined and richer in meaning. In short, through interaction with the opposing army, the army as a whole becomes more clearly reflected in the mind of each of its members, its self-consciousness is clarified and enriched. In a similar way, intercourse and rivalry between the various regiments greatly promotes the growth of the self-knowledge and self-sentiment of each of these lesser groups. A standing army inevitably possesses a wealth of traditions, habits and customs, over and above its formal organisation, and these play an important part in promoting the smooth working of the whole organism; the lack of these is one of the chief difficulties in the way of the creation of a new army, as was vividly illustrated in the making of the "Kitchener army" during the Great War. The customs of the various officers' messes were but a small part of this mass of custom which does so much to bind the whole army together.

An army obviously possesses organisation, generally in a very high degree. The formal continuity of its existence enables the organisation impressed upon it by external authority to acquire all the strength that custom alone can give; while its material continuity enables its organisation to generate, in the individual soldiers, habits through which the inferior members are raised, as regards the moral qualities required for efficiency in the field, towards the level of the best.

The organisation of the whole army has two aspects and two main functions; the one is executive, the securing of the co-ordination of action of the parts in the carrying out of the common plan; the other is recipient and deliberative, the co-ordination of the data supplied by the parts through deliberation upon which the choice of means is arrived at. Deliberation and choice of means are carried out by the commander-in-chief and his staff, the persons who have shown themselves best able to execute this part of the army's task. It is important to note that, in the case of such an army as we are considering, the private soldier in the ranks remains a free agent performing truly volitional actions; that he in no sense becomes a mechanical agent or one acting through enforced or habitual obedience merely. He wills the common end; and, believing that the choice of means to that end is best effected by the appropriate part of the whole organisation, he accepts the means chosen, makes of them his proximate end, and wills them.

This is the essential character of the effective organisation of any human group; it secures that while the common end of collective action is willed by all, the choice of means is left to those best qualified and in the best position for deliberation and choice; and it secures that co-ordination of the voluntary actions of the parts which brings about the common end by the means so chosen. In this way the collective actions of the well-organised group, instead of being, like those of the simple crowd, merely impulsive or instinctive actions, implying a degree of intelligence and of morality far inferior to that of the average individual of the crowd, become truly volitional actions expressive of a degree of intelligence and morality much higher than that of the average member of the group: *i. e.*, the whole is raised above the level of its average member; and even, by reason of exaltation of emotion and organised co-operation in deliberation, above that of its highest members.

Here we must consider a little more fully the nature of the collective or general will, a subject that has figured largely in the discussions of political philosophers on the nature of the State. Rousseau wrote—"There is often a great difference between the will of all and the general will; the latter looks only to the common interest; the former looks to private interest, and is nothing but a series of individual wills; but take away from these same wills the plus and minus that cancel one another and there remains, as the sum of the differences, the general will." "Sovereignty is only the exercise of the general will." By this he seems to mean that a certain number of men will the general good, while many will only their private goods and that while the latter neutralize one another, as regards their effects on the general interest, the former co-operate and so form an effective force to promote the general good. This doctrine was an approximation towards the truth, though like all Rousseau's social speculations, his handling of it was vitiated by his false psychology, which set out from the fiction of man as an independent purely self-contained and self-determining absolute individual. Later writers do not seem to have improved upon Rousseau's doctrine of the general will to any great extent.

The problem of the general will, like all problems of collective psychology, becomes extremely complex when we consider the life of nations; and it is, therefore, important to make ourselves clear as to the nature of collective volition by consideration of the relatively simple case of a patriot army. It is of course impossible to arrive at a clear notion of collective volition, until individual volition has been clearly defined and the nature of the distinction between it, on the one hand, and mere impulsive action, desire, and simple conflict of desires, on the other hand, has been made clear. The lack of such clear notions and adequate definitions has rendered much of the discussion of this topic by political philosophers sterile and obscure. In the light of the conclusions reached in my chapter on individual volition,[1] the question of the nature of collective volition presents little difficulty. It was found that volition may be defined and adequately marked off from the simpler modes of conation by saying that it is the re-inforcement of any impulse or conation by one excited within the system of the self-regarding sentiment. And in an earlier chapter[2] it was shown how the self-regarding sentiment may become extended to other objects than the individual self, to all objects with which the self identifies itself, which are regarded as belonging to the self or as part of the wider self. This extension depends largely on the fact that others identify us with such an object, so that we feel ourselves to be an object of all the regards and attitudes and actions of others directed towards that object, and are emotionally affected by them in the same ways that we are affected by similar regards, attitudes, and actions directed towards us individually. It was shown also that such a sentiment may become wider and emotionally richer than the purely self-regarding sentiment, through fusing with a sentiment of love for the object that

[1] *Social Psychology*, Chapter IX. [2] *Op. cit.*, Chapter VII.

has grown up independently. These facts were illustrated by consideration of the parental sentiment for the child, which, it was said, has commonly this twofold character and source, being formed by the compounding of the self-regarding sentiment with the sentiment of love, of which the dominant disposition is that of the tender or protective instinct.

In a similar way a similarly complex sentiment may become organised about the idea of one's family, or of any still larger group having continuity of existence of which one becomes a member. In the case of the patriot's sentiment for his country or nation, the self-regarding sentiment and the sentiment of love may be from the first combined in the patriotic sentiment; since he knows himself to be a part of the whole from the time that an idea of the whole first takes shape in his mind.

In this respect the case of the soldier in a patriot army is relatively simple. As a boy he may have acquired a sentiment of loving admiration for the army; and, when he becomes a member of it, the dispositions that enter into the constitution of his self-regarding sentiment become incorporated with this previously existing sentiment, so that the reputation of the army becomes as important to him as his own; praise and approval of it become for him objects of desire and sources of elation; disapproval and blame of it, or the prospect of them, affect him as painfully as if directed to himself individually, fill him with shame and mortification.

A similar complex sentiment, the sentiment of patriotism, becomes organised about the idea of his country as a whole; and, when war breaks out and the army is pitted against that of another nation, while the eyes of the whole world are turned upon it, it becomes the representative of the nation and the special object of the patriotic sentiment, which thus adds its strength to that of the more special sentiment of the soldier for the army.[1] When, then, the patriot army takes the field, it is capable of collective volition in virtue of the existence of this sentiment in the minds of all its members. The soldiers of a purely mercenary army are moved by the desire of individual glory, of increased pay, of loot, by the habit of obedience and collective movement acquired by prolonged drilling, by the pugnacious impulse, by the desire of self-preservation; and they may be led on to greater exertions by the influence of an admired captain. But such an army is incapable of collective volition, because no sentiment for the army as a whole is common to all its members. The soldiers of the patriot army on the other hand may act from all the individual motives enumerated above; but all alike are capable also of being stirred by a common motive, a desire excited within the collective self-regarding sentiment, the common sentiment for the army; and this, adding itself to whatever individual motives are operative, converts their desires into collective resolutions and renders their actions the expressions of a collective volition.

Each soldier of the mercenary army may desire that his side shall win the battle and may resolve that he will do his best to bring victory to his side, and he may perform many truly volitional actions; and, in so far as the actions of the army express these individual volitions towards a common result, they are the expressions of the "will of all," but not of the collective will; because these volitions, though they are directed to the one common end, spring from diverse motives and are individual volitions.

The essence of collective volition is, then, not merely the

[1] One great difference between the professional army such as that of England and the citizen armies of Europe, consists in the fact that the special sentiment for the army is stronger in the former; the more general patriotic sentiment, in the rank and file of the latter; though in the regular officers of the continental army the sentiment for the army itself is no doubt usually the stronger.

direction of the wills of all to the same end, but the motivation of the wills of all members of a group by impulses awakened within the common sentiment for the whole of which they are the parts. It is the extension of the self-regarding sentiment of each member of the group to the group as a whole that binds the group together and renders it a collective individual capable of collective volition.

The facts may be illustrated more concretely by taking a still simpler example of collective volition. Consider the case of a regiment in battle commanded to occupy a certain hilltop in face of fierce opposition. If the regiment is one to which the self-regarding sentiment of each member has become extended, the soldiers may be animated individually by the pugnacious impulse and by the desire of individual glory, but they are moved also by the common desire to show what the regiment can do, to sustain its glorious reputation; they resolve that we, the regiment, will accomplish this feat. As they charge up the hill, the hail of bullets decimates their ranks and they waver, the impulse of fear checking their onward rush; if then their officer appeals to the common sentiment, each man feels the answering impulse; and this is strengthened by the cheer which shows him that the same impulse rises in all his comrades; and so this impulse, awakened within the collective self-regarding sentiment and strengthened by sympathetic induction from all to each, comes to the support of the pugnacious impulse or whatever other motives sustain each man, enables these to triumph over the impulse to flight, and sweeps them all on to gain their object by truly collective volitional effort. If, on the other hand, the men of the regiment have no such common sentiment, then, when the advancing line wavers, the onward impulsion checked by the impulse of retreat, there is no possibility of arousing a collective volition; the regiment, which from the first was a crowd organised only by external authority and the habits created by it, acts as a crowd and yields to the rising impulse of the emotion of fear, which, becoming intensified by induction from man to man, rises to a panic; and the regiment is routed.

We may distinguish, then, five modes of conation which will carry all the members of a group towards a common object, five levels of collective action.

Let the group be a body of men on a road leading across a wilderness to a certain walled city. A sudden threat of danger from a band of robbers or from wild beasts may send them all flying in panic towards the city gate. That is a purely impulsive collective action. It is not merely a sum of individual actions, because the fear and, therefore, the impulse to flight of each man is intensified by the influence of his fellows.

Secondly, let them be a band of pilgrims, fortuitously congregated, each of whom has resolved to reach the city for his own private purposes. The whole body moves on steadily, each member aided in maintaining his resolution in face of difficulties by the presence of the rest and the spectacle of their resolute efforts. Here there is a certain collectivity of action; the individual wills are strengthened by the community of purpose. But the arrival of the band is not due to collective volition; nor can it properly be said to be due to the will of all; for each member cares nothing for the arrival of the band as a whole; he desires and wills only his own arrival.

Thirdly, let each member of the band be aware that, at any point of the road, robbers may oppose the passage of any individual or of any company not sufficiently strong to force its way through. Each member will then desire that the whole band shall cohere and shall reach the city, and the actions of the group will display a higher degree of co-operation and collective efficiency than in the former cases; but the successful passage of the band will be desired by each member simply in order that his own safe arrival may be secured. There is direction of all wills towards the production of the one result, the success of the whole band; but this is not truly collective volition because the motives are private and individual and diverse.

Fourthly, let the band be an army of crusaders, a motley throng of heterogeneous elements of various nationalities, united by one common purpose, the capture of the city, but having no sentiment for the army. In this case all members not only will the same collective action and desire the same end of that action, but they have similar motives arising from their sentiment for the city or that which it contains. Still their combined actions are not the issue of a collective volition in the full and proper sense of the words, but of a coincidental conjunction of individual volitions. They might perhaps be said to be the expression of the general will; and by giving that meaning to the term "general will," while reserving the expression collective will or volition for the type of case illustrated by our next instance, we may usefully differentiate the two expressions.

Lastly, let the band approaching the city be an army of crusaders of one nationality, and let us suppose that this army has enjoyed a considerable continuity of existence and that in the mind of each member the self-regarding sentiment has become extended to the army as a whole, so that, as we say, each one identifies himself with it and prizes its reputation and desires its success as an end in itself. Such a sentiment would be greatly developed and strengthened by rivalry in deeds of arms with a second crusading army. Each member of this army would have the same motives for capturing the city as those of the army of our last instance; but, in addition to these motives, there would be awakened within the extended self-regarding sentiment of each man an impulse to assert the power, to sustain the glory of the army; and this, adding its force to those other motives, would enable them to triumph over all conflicting tendencies and render the resolution of the army to capture the city a true collective volition; so that the army might properly be said to possess and to exercise a general or collective will.

This distinction between the will of all and the collective will, which we have considered at some length, may seem to be of slight importance in the instances chosen. But it becomes of the greatest importance when we have to consider the life of a nation or other enduring community. The power of truly collective volition is no small advantage to any body of fighting men and receives practical recognition from experienced captains.

The importance of these different types of volition was abundantly illustrated by the incidents of the Boer war and of the Russo-Japanese war. That the success of its undertaking shall be strongly willed by all is perhaps the most important factor contributing to the success of an army; and if also the army exercises a true collective volition, in the sense defined above, it becomes irresistible. Though it is questionable whether the Boer armies can be said to have exercised a collective volition, it is at least certain that individually the Boers strongly willed their common end, the defeat of the British. On the other hand the British armies were defective in these respects. The motives of those who fought in the British armies against the Boers were very diverse. The pay of the regulars, the five shillings a day of the volunteers, the desire to live for a time an adventurous exciting life, the desire to get home again on the sick-list as soon as possible, the desire for personal distinction; all these and other motives were in many minds mixed in various proportions with the desire to assert the supremacy of the British rule and support the honour of the flag. This difference between the Boer and British armies was undoubtedly a main cause of many of the surprising successes of the former. In the Russo-Japanese war the opposed armies probably differed even more widely in this respect. The Japanese soldiers not only willed intensely the common end, but their armies would appear to have exercised truly collective volition. Many of the several regiments also, being recruited on the territorial system, were animated by collective sentiments rooted in local patriotism. The Russian armies on the other hand were largely composed of peasants drawn from widely separated regions of the Russian empire, knowing little or nothing of the grounds of quarrel or of the ends to be achieved by their efforts, caring nothing individually for those ends, and having but little patriotic sentiment and still less sentiment for the army.

It would, then, be a grave mistake to infer from the course of events in these two wars that the British soldier was individually inferior to the Boer, or the Russian to the Japanese; in both cases the principal psychological condition of successful collective action—namely, a common end intensely desired and strongly willed, individually or collectively—was present in high degree on the one side, because the preservation of the national existence was the end in view; while it was lacking or comparatively deficient on the other side. As Sir Ian Hamilton, a close observer of both these wars, has said—"the army that will not surrender under any circumstances will always vanquish the army whose units are prepared to do so under sufficient pressure."

The same considerations afford an explanation of a peculiarity of Russian armies which has often been noted in previous wars, and which was very conspicuous in the late wars; namely, their weakness in attack and their great strength when on the defensive. For, in attacking, a Russian army is in the main merely obeying the will of the commander-in-chief in virtue of custom, habit, and a form of strong collective suggestion; but in retreat and on the defensive, each man's action becomes truly volitional, all are animated by a common purpose, and all will the same end, the safety of the whole with which that of each member is bound up.

The psychology of a patriot army is peculiarly simplified as compared with that of most other large human groups, by two conditions; on the one hand, the restriction of the intellectual processes, by which the large means for the pursuit of the common end are chosen, to one or a few minds only; on the other hand, the definiteness and singleness of its purpose and the presence of this clear and strong purpose in the minds of all.

Other groups that enjoy in some degree the latter condition of simplicity of collective mental life are associations voluntarily formed and organised for the attainment of some single well-defined end. In them the former condition is generally completely lacking and the deliberative processes, by which their means are chosen, are apt to be very complex and ineffective, owing to lack of customary organisation. Such associations illustrate more clearly than any other groups the part played by the idea of the whole in the minds of the individuals in constituting and maintaining the whole. A desire or purpose being present in many minds, the idea of the association arises in some one or more of them, and, being communicated to others, becomes the immediate instrument through which the association is called into being; and only so long as this idea of the whole as an instrument for attaining the common end persists in the minds of the individuals does the association continue to exist. In this respect such an association is at the opposite end of the scale from the fortuitous crowd, which owes its existence to the accidents of time and place merely. Human groups of other kinds owe their existence in various proportions to these two conditions; such groups, for example, as are constituted by the members of a church, of a university or a school, of a profession or a township. Others, such as nations, owe their inception to the accidents of time and place, to physical boundaries and climatic conditions; and, in the course of their evolution, become more and more dependent for their existence on the idea of the whole and the sentiment organised about it in the minds of their members; and they may, like the Jewish people, arrive in the course of time at complete dependence on the latter condition.

The life of an army illustrates better than that of any other group the influence of leadership. That great strategists and skilful tacticians perform intellectual services of immeasurable importance for the common end of the army goes without saying. But the moral influence of leadership is more subtle in its workings, and is perhaps less generally recognised in all its complexity and scope. It is well known that such commanders as Napoleon inspired unlimited confidence and enthusiasm in the veteran armies that had made many campaigns under their leadership. Yet in the Great War, in which the British armies were, in its later stages, composed so largely of new recruits, the same influence was perceptible. Both the British and the French armies were very fortunate in having in supreme command men in whom the common soldier felt confidence. The solidity, the justice, the calm resolution of Marshal Joffre were felt throughout the French army in the early days of the war to be the one certain and fixed point in a crumbling universe. "Il est solide, le Père Joffre" was repeated by thousands who, remembering the disaster of 1870, were inclined to suspect treachery and weakness on every hand. And the genius of Marshal Foch and of other brilliant generals was a main source of the astonishing dogged resolution with which the French armies, in spite of their terrible losses, sustained the prolonged agony. The British army also was fortunate in having in Field-Marshal Haig a man at its head who was felt to be above all things resolute and calm and just; and, when the British armies in France were placed under the supreme control of Foch, it was generally felt throughout the ranks that this would not only give unity of control and purpose, but also supply that touch of genius which perhaps had been lacking in British strategy.

But it was not only the supreme command that exercised this influence over the minds of all ranks. At every level, confidence in the leadership was of supreme importance. The character and talents of each general and colonel, of each captain, lieutenant, sergeant, and corporal, made themselves felt by all under their control; felt not only individually but corporately and collectively. The whole area under the command of any particular general might be seen to reflect and to express in some degree his attributes. The reputations of the higher officers filtered down through the ranks in an astonishingly rapid and accurate manner; perhaps owing largely to the fact that these armies, in a degree unknown before, were composed of men accustomed to read and to think and to discuss and criticise the conduct of affairs. If the German higher command had been exercised from the first by a man who inspired the just confidence that was felt in the old Field-Marshal v. Moltke by the Prussian armies of 1870, it is probable that the issue of the Great War would have been fatally different.

The moral effects of good leadership are, perhaps, of more importance to an army than its intellectual qualities, especially in a prolonged struggle; and these work throughout the mass of men by subtle processes of suggestion and emotional contagion rather than by any process of purely intellectual appreciation. And the whole organisation of any wisely directed army is designed to render as effective as possible these processes by which the influence of leaders is diffused through the whole.

CHAPTER IV

The Group Spirit

IN considering the mental life of a patriot army, as the type of a highly organised group, we saw that *group self-consciousness* is a factor of very great importance—that it is a principal condition of the elevation of its collective mental life and behaviour above the level of the merely impulsive violence and unreasoning fickleness of the mob.

This self-consciousness of the group is the essential condition of all higher group life; we must therefore study it more nearly as it is manifested in groups of various types. It is unfortunate that our language has no word that accurately translates the French expression, *esprit de corps;* for this conveys exactly the conception that we are examining. I propose to use the term "group spirit" as the equivalent of the French expression, the frequent use of which in English speech and writing sufficiently justifies the attempt to specialise this compound word for psychological purposes.

We have seen that, in virtue of the sentiment developed about the idea of the army, all its members exhibit *group loyalty;* it is only as the sentiment develops about the idea that this idea of the whole, present to the mind of each member, becomes a power which can hold the whole group together, in spite of all physical and moral difficulties. We see this if we reflect how armies of mercenaries, in which this collective sentiment is lacking or rudimentary only, are apt to dissolve and fade away by desertion as soon as serious difficulties are encountered.

The importance of the collective idea and sentiment appears still more clearly, when we reflect on the type of army which has generally proved the most efficient of all —namely, an army of volunteers banded together to achieve some particular end. Such an army (for example the army of Garibaldi) owes its existence to the operation of this idea in the minds of all. The idea of the army is formed in the mind perhaps of one only (Garibaldi); he communicates it to others, who accept it as a means to the end desired by all of them individually. The idea of the whole thus operates to create the group, to bring it into existence; and then, as the idea is realised, it becomes more definite, of richer and more exact meaning; the collective sentiment grows up about it, and habit and formal organisation begin to aid in holding the group together; yet still the idea of the whole remains constitutive of the whole.

Any group that owes its creation and its continued existence to the collective idea may be regarded from the psychological standpoint as of the highest type; while a fortuitously gathered crowd that owes its existence to accidents of time and place and has the barest minimum of group self-consciousness is of the lowest type. Every other form of association or of human group may be regarded as occupying a position in a scale between these extreme types; according to the relative predominance of the mental or the physical conditions of its origin and continuance, that is to say, according to the degree in which its existence is teleologically or mechanically determined.

The group spirit, the idea of the group with the sentiment of devotion to the group developed in the minds of all its members, not only serves as a bond that holds the group together or even creates it, but, as we saw in the case of the patriot army, it renders possible truly collective volition; this in turn renders the actions of the group much more resolute and effective than they could be, so long as its actions proceed merely from the presence of an impulse common to all members, or from the strictly individual volitions of all, even though these be directed to one common end.

Again, the group spirit plays an important part in raising the intellectual level of the group; for it leads each member deliberately to subordinate his own judgment and opinion to that of the whole; and, in any properly organised group, this collective opinion will be superior to that of the average individual, because in its formation the best minds, acting upon the fullest knowledge to the gathering of which all may contribute, will be of predominant influence. Each member, then, willing the common end, accepts the means chosen by the organised collective deliberation, and, in executing the actions prescribed for him, makes them his own immediate ends and truly wills them for the sake of the whole, not executing them in the spirit of merely mechanical unintelligent obedience or even of reluctance.

In a similar way the group spirit aids in raising the moral level of an army. The organised whole embodies certain traditional sentiments, especially sentiments of admiration for certain moral qualities, courage, endurance, trustworthiness, and cheerful obedience; and these sentiments, permeating the whole, are impressed upon every member, especially new members, by way of mass suggestion and sympathetic contagion; every new recruit finds that his comrades accept without question these traditional moral sentiments and confidently express moral judgments upon conduct and character in accordance with them, and that they also display the corresponding emotional reactions towards acts; that is to say, they express in verbal judgments and in emotional reactions their scorn for treachery or cowardice, their admiration for courageous self-sacrifice and devotion to duty. The recruit quickly shares by contagion these moral emotions and soon finds his judgment determined to share these opinions by the weight of mass suggestion; for these moral propositions come to him with all the irresistible force of opinion held by the group and expressed by its unanimous voice; and this force is not merely the force derived from numbers, but is also the force of the prestige accumulated by the whole group, the prestige of old and well-tried tradition, the prestige of age; and the more fully the consciousness of the whole group is present to the mind of each member, the more effectively will the whole impress its moral precepts upon each.

And the organisation of the army renders it possible for the leaders to influence and to mould the form of these moral opinions and sentiments. Thus Lord Kitchener, by issuing his exhortation to the British Army on its departure to France, did undoubtedly exert a considerable influence towards raising the moral level of the whole force, because he strengthened the influence of those who were already of his way of thinking against the influence of those whose sentiments and habits and opinions made in the opposite direction. His great prestige, which was of a double kind, both personal and due to his high office, enabled him to do this. In the same way, every officer in a less degree can do something to raise the moral level of the men under his command. Thus, then, the organisation of the whole group, with its hierarchy of offices which confer prestige, gives those who hold these higher offices the opportunity to raise the moral level of all members.

Of course, if those who occupy these positions of prestige feel no responsibility of this sort and make no effort to exert such influence, but rather aim at striking terror in the foe at all costs, if they countenance acts of savagery such as the destruction of cities, looting, and rapine, if they publicly instruct their soldiers to behave as Huns or savages; then the organisation of the army works in the opposite way—namely, to degrade all members below their normal individual level, rather than to raise them above it; and then we hear of acts of brutality on the part of the rank and file which are almost incredible.

But the main point to be insisted on here is that the raising of the moral level is not effected only by example, suggestion, and emotional contagion, spreading from those in the positions of prestige; that, where the group spirit exists, those enjoying prestige can, if they wish, greatly promote the end of raising the moral tone of the whole by appealing to that group spirit; as when Lord Kitchener asked the men to obey his injunctions for the sake of the honour of the British army.

And the group spirit not only yields this direct response to moral exhortation; it operates in another no less important manner. Each member of a group pervaded by the collective sentiment, such as a well-organised army of high traditions, becomes in a special sense his brother's keeper. Each feels an interest in the conduct of every other member, because the conduct of each affects the reputation of the whole; each man, therefore, punishes bad conduct of any fellow-soldier by scorn and by withdrawal of sympathy and companionship; and each one rewards with praise and admiration the conduct that conforms to the standards demanded and admired. And so each member acts always under the jealous eyes of all his fellows, under the threat of general disapprobation, contempt, and moral isolation for bad conduct; under the promise of general approval and admiration for any act of special excellence.

The development of the group spirit, with the appropriate sentiment of attachment or devotion to the whole and therefore also to its parts, is the essence of the higher form of military discipline. There is a lower form of discipline which aims only at rendering each man perfectly subservient to his officers and trained to respond promptly and invariably, in precise, semi-mechanical, habitual fashion, to every word of command. But even the drill and the system of penalties and minute supervision, which are the means chosen to bring about this result, cannot fail to achieve certain effects on a higher moral and intellectual level than the mere formation of bodily habits of response. By rendering each soldier apt and exact in his response to commands, they enable each one to foresee the actions of his fellows in all ordinary circumstances, and therefore to rely upon that co-operation towards the common end, be it merely a turning movement on the drill ground or the winning of a battle, which is the essential aim and justification of all group life.

The group spirit, involving knowledge of the group as such, some idea of the group, and some sentiment of devotion or attachment to the group, is then the essential condition of all developed collective life, and of all effective collective action; but it is by no means confined to highly developed human associations of a voluntary kind.

Whether the group spirit is possessed in any degree by animal societies is a very difficult question. We certainly do not need to postulate it in order to account for the existence of more or less enduring associations of animals; just as we do not need to postulate it to account for the coming together of any fortuitous human mob. Even in such animal societies as those of the ants and bees, its presence, though often asserted, seems to be highly questionable. When we observe the division of labour that characterises the hive, how some bees ventilate, some build the comb, some feed the larvæ and so on; and especially when we hear that the departure of a swarm from the hive is preceded by the explorations of a small number which seek a suitable place for the new home of the swarm and then guide it to the chosen spot, it seems difficult to deny that some idea of the community and its needs is present to the minds of its members. But we know so little as yet of the limits of purely instinctive behaviour (and by that I mean immediate reactions upon sense-perceptions determined by the innate constitution) that it would be rash to make any such inference. The same may be said of associations of birds or mammals, in which division of labour is frequently displayed; when, for example, it is found that one or more sentinels constantly keep watch while a flock or herd feeds or rests, as is reported of many gregarious species.

But, however it may be with animal societies, we may confidently assert that the group spirit has played an important part in the lives of all enduring human groups, from the most primitive ages onwards.

It has even been maintained with some plausibility that group self-consciousness preceded individual self-consciousness in the course of the evolution of the human mind. That again is, it seems to me, a proposition which cannot be substantiated. But it is, I think, true to say that the two kinds of self-consciousness must have been achieved by parallel processes, which constantly reacted upon one another in reciprocal promotion.

In the lives of the humblest savages the group spirit plays an immensely important part. It is the rule that a savage is born into a small closed community. Such a community generally has its own locality within which it remains, even if nomadic; and, if settled, it wholly lives in a village, widely separated in space from all others. In this small community the child grows up, becoming more or less intimately acquainted with every member of it, and having practically no intercourse with any other persons.

Throughout his childhood he learns its laws and traditions, becomes acutely aware of its public opinion, and finds his welfare absolutely bound up with that of the village community. He cannot leave it if he would; the only alternative open to him is to become an outcast, as which he would very soon succumb in the struggle for life. There is nothing comparable with this in our complex civilised societies. The nearest parallel to it is the case of the young child growing up in a peculiarly secluded family isolated in the depths of the country.

This restriction of the intercourse of the young savage to the members of his own small society and his absolute dependence upon it for all that makes his survival possible would in themselves suffice to develop his group-consciousness in a high degree. But two other conditions, well-nigh universal in savage life, tend strongly towards the same result.

When the young savage begins to come into contact with persons other than those of his own group, he learns to know them, not as individuals, John Smith or Tom Brown, but as men of such or such a group; and he himself is known to them as a man of his group, as representing his group, his village community, tribe, or what not; and he displays usually some mark or marks of his group, either in dress or ornament or speech.

The other great condition of the development of the group spirit in primitive societies is the general recognition of communal responsibility. This no doubt is largely the result of the two conditions previously mentioned, especially of the recognition of an individual by members of other groups as merely a representative of his group, rather than as an individual, and of the fact that his deeds, or those of any one of his fellows, determine the attitudes of other groups towards his group as a whole. But the influence of the principle of communal responsibility, thus established, becomes immensely strengthened by its recognition in a number of superstitious and religious observances. The savage lives, generally speaking, bound hand and foot by *tabus* and precise prescriptions of behaviour for all ordinary situations; and the breach of any one of these by any member of the community is held to bring down misfortune or punishment on the whole group; so far is this principle carried, that the breach of custom by some individual is confidently inferred from the incidence of any communal misfortune.[1]

The recognition of communal responsibility is the great conservator of savage society and customary law, the very root and stem of all savage morality; it is the effective moral sanction without which the superstitious and religious sanctions would be of little effect. By its means, the idea of the community is constantly obtruded on the consciousness of the individual. Through it he is constantly led, or forced, to control his individualistic impulses and to undertake action with regard to the welfare of the group rather than to his own private interest. Through it the tendency of each to identify himself and each of his fellows with the whole group is constantly fostered; because it identifies their interests.

We may then say that, just as the direct induction of emotion and impulse by sense-perception of their bodily expressions is the cement of animal societies, so group self-consciousness is the cement and harmonising principle of primitive human societies.

And the group spirit is not only highly effective in promoting the life and welfare of the group; it is also the source of peculiar satisfactions. The individual revels in his group-consciousness; hence the principle is apt to run riot in savage societies, and we find that in very many

[1] Cp. *The Pagan Tribes of Borneo*, by Ch. Hose and W. McDougall, London, 1912.

parts of the world a great variety of complex forms of association is maintained, beside the primary and fundamental form of association of the village community or nomadic band (the kinship or subsistence group), apparently for no other reason than the attainment and intensification of the satisfactions of the group spirit. Hence, among peoples so low in the scale of savagery as the Australians, we find a most complex system of grouping cutting across the subsistence grouping; hence totem clans and phratries, exogamous groups, secret societies, initiation ceremonies.

I lay stress on the satisfaction which group self-consciousness brings as a condition or cause of these complexities of savage society, because, I think, it has been unduly neglected as a socialising factor and a determinant of the forms of association. If we ask—What are the sources of this satisfaction?—we may find two answers. First, the consciousness of the group and of oneself as a member of it brings a sense of power and security, an assurance of sympathy and co-operation, a moral and physical support without which man can hardly face the world. In a thousand situations it is a source of settled opinions and of definite guidance of conduct which obviates the most uncomfortable and difficult necessity of exerting independent judgment and making up one's own mind. And in many such situations, not only does the savage find a definite code prescribed for his guidance, but he shares the collective emotion and feels the collective impulse that carries him on to action without hesitation or timidity.

Secondly, we may, I think, go back to a very fundamental principle of instinctive life, the principle, namely, that, in gregarious animals, the satisfaction of the gregarious impulse is greater or more complete the more nearly alike are the individuals congregated together. This seems to be true of the animals, but it is true in a higher degree of man; and, in proportion as his mind becomes more specialised and refined, the more exacting is he in this respect. To the uncultivated any society is better than none; but in the cultivated classes we become extraordinarily exacting; we find the gregarious satisfaction in our own peculiar set only—a process carried furthest, perhaps, in university circles. In savage life this shows itself in practices which accentuate the likeness of members of a group and mark it off more distinctly from other groups—for example, totems, peculiarities of dress, ornaments and ceremonies; things which are closely paralleled by the clubs, blazers, colours, cries, and so forth of our undergraduate communities.

The life of the savage, then, is in general dominated by that of the group; and this domination is not effected by physical force or compulsion (save in exceptional instances) but by the group spirit which is inevitably developed in the mind of the savage child by the material circumstances of his life and by the traditions, especially the superstitious and religious traditions, of his community. Such group self-consciousness is the principal moralising influence, and to this influence is due in the main the fact that savages conform so strictly to their accepted moral codes.

Group self-consciousness in savage communities brings then, I suggest, two great advantages which account for the spontaneous development and persistence among so many savage peoples of what, from a narrowly utilitarian point of view, might seem to be an excess of group organisation, such as the totemic systems of the Australians and of the American Indians;—namely, firstly, the moralising influences of the group spirit; secondly, the satisfactions or enjoyments immediately accruing to every participant in active group life.

And these two advantages, being in some degree appreciated, lead to a deliberate cultivation of group life for the securing of them in higher measure. The cultivation of group life shows itself in the many varieties of grouping on a purely artificial basis and in the practice of rites and ceremonies, especially dances, often accompanied by song and other music. There is nothing that so intensifies group-consciousness, at the cost of consciousness of individuality, as ceremonial dancing and singing; especially when the dance consists of a series of extravagant bizarre movements, executed by every member of the group in unison, the series of movements being at the same time peculiar to the particular group that practises them and symbolical of the peculiar functions or properties claimed by the group. Many savage dances have these characters in perfection; as, for example, those of the Murray islanders of Torres Straits, where, as I have witnessed, the several totemic groups—the dog-men, the pigeon-men, the shark-men, and other such groups—continue, in spite of the partial destruction by missionaries of their totemistic beliefs, to revel in night-long gatherings, at which each group in turn mimics, in fantastic dances and with solemn delight, the movements of its totem animal.

The importance of group-consciousness in savage life has been recently much insisted on by some anthropologists, and indeed, in my view, overstated. Cornford[1] writes, "When the totem-clan meets to hold its peculiar dance, to work itself up till it feels the pulsing of its common life through all its members, such nascent sense of individuality as a savage may have—it is always very faint—is merged and lost; his consciousness is filled with a sense of sympathetic activity. The group is now feeling and acting as one soul, with a total force much greater than any of its members could exercise in isolation. The individual is lost, 'beside himself,' in one of those states

[1] *From Religion to Philosophy*, p. 77.

of contagious enthusiasm in which it is well known that men become capable of feats which far outrange their normal powers." And again "Over and above their individual experience, all the members of the group alike partake of what has been called the collective consciousness of the group as a whole. Unlike their private experience, this pervading consciousness is the same in all, consisting in those epidemic or infectious states of feeling above described, which, at times when the common functions are being exercised, invade the whole field of mentality, and submerge the individual areas. To this group-consciousness belong also, from the first moment of their appearance all representations which are collective, a class in which all religious representations are included. These likewise are diffused over the whole mentality of the group, and are identical in all its members. . . . The collective consciousness is thus superindividual. It resides, of course, in the individuals composing the group. There is nowhere else for it to exist, but it resides in all of them together and not completely in any one of them. It is both in myself and yet not myself. It occupies a certain part of my mind and yet it stretches beyond and outside me to the limits of my group. And since I am only a small part of my group, there is much more of it outside me than inside. Its force therefore is much greater than my individual force, and the more primitive I am the greater this preponderance will be. Here, then, there exists in the world a power which is much greater than any individual's—superindividual, that is to say superhuman."

"Because this force is continuous with my own consciousness, it is, as it were, a reservoir to which I have access, and from which I can absorb superhuman power to reinforce and enhance my own. This is its positive aspect —in so far as this power is not myself and greater than myself, it is a moral and restraining force, which can and does impose upon the individual the necessity of observing the uniform behaviour of the group." This writer makes group-consciousness the source of both morality and religion. "The collective consciousness is also immanent in the individual himself, forming within him that unreasoning impulse called conscience, which like a traitor within the gates, acknowledges from within the obligation to obey that other and much larger part of the collective consciousness which lies outside. Small wonder that obedience is absolute in primitive man, whose individuality is still restricted to a comparatively small field, while all the higher levels of mentality are occupied by this overpowering force."[1] The first religious idea is that of "this collective consciousness, the only moral power which can come to be felt as imposed from without." And Cornford goes yet further and makes of the group self-consciousness the source of magic as well as of religion and morality. This primary reservoir of superindividual power splits, he says, into two pools, human and non-human; the former is magic power, the latter is divine power.

On this I would comment as follows. Although Cornford is right in insisting upon the large influence of group-consciousness, he is wrong, I think, in underestimating individuality. He does not go so far as some writers who suggest that group self-consciousness actually precedes individual self-consciousness, but he says of individual self-consciousness that it is but very faint in savages. I am more inclined to agree with Lotze, who in a famous passage asserted that even the crushed worm is in an obscure way aware of itself and its pain as set over against the world. Many facts of savage behaviour forbid us to accept the extreme view that denies them individual self-consciousness—individual names, secret names, private property, private rites, religious and magical, individual

[1] *Op. cit.*, p. 82.

revenge, jealousy, running *amok*, leadership, self-assertion, pride, vanity, competition in games of skill and in technical and artistic achievement. The flourishing of these and many other such things in primitive communities reveals clearly enough to the unbiassed observer in the field the effective presence of individual self-consciousness in the savage mind. In this connection I may refer to two pieces of evidence bearing very directly on the question reported by Dr. C. Hose and myself.[1] Among the Sea-Dayaks or Ibans of Borneo we discovered the prevalence of the belief in the "nyarong" or private "spirit helper," some spiritual or animated individual power which a fortunate individual here and there finds reason to believe is attached to him personally for his guidance and help in all difficult situations. His belief in this personal helper and the rites by aid of which he communicates with it are kept secret from his fellows; so that it was only after long and intimate acquaintance with these people that Dr. Hose began to suspect the existence of this peculiarly individualistic belief.[1] In the same volumes we have described the Punans of Borneo, a people whose mode of life is in every respect extremely primitive. In this respect they are perhaps unequalled by any other existing people. Yet no one who is acquainted with these amiable folk could doubt that, although their group-consciousness is highly developed, they enjoy also a well-developed individual self-consciousness. How otherwise can we interpret the fact that a Punan who suffers malicious injury from a member of another tribe will nurse his vengeful feeling for an indefinite period and, after the lapse of years, will find an opportunity to bring down his enemy secretly with blowpipe and poisoned dart?

With Mr. Cornford's view of the part played by the

[1] Cp. *The Pagan Tribes of Borneo*, by Ch. Hose and W. McDougall. London, 1912.

group spirit in moralising conduct I agree. I agree also that it is the collective life or mind that develops religion and in part magic; but in my view Cornford attributes to the savage far too much reflective theorising; he represents him as formulating a theory of the collective consciousness which is really almost identical with the interesting speculation of M. Lévy Bruhl presently to be noticed; and he regards his conduct, his religious and magical practices, as guided by these theories. But that is to reverse the true order of things—to make theory precede practice; whereas in reality, especially in religion and magic, practice has everywhere preceded theory, often, as in this case, by thousands of generations.[1]

It is true that the savage often behaves as though he held this theory of the collective consciousness as a field of force in which he participates; that his conduct seems to require such a theory for its rational justification. But it by no means follows that he has formulated any theory at all. What the savage is conscious of is, not a collective consciousness as a mysterious superhuman power, but the group itself, the group of concrete embodied fellow-men. He behaves and feels as he does, because participation in the life of the group directly modifies his individual tendencies and directly evokes these feelings and actions; he does not discover, or seek, any theory by which to explain them. Still less is it true that he performs these actions because he has formulated a theory of a collective consciousness.

Mr. Cornford regards the savage idea of a collective consciousness as the germ of the idea of divine power or of God. Now this is connected with the question of animism, preanimism, and dynanimism. It may be true that the notion of *mana* is the common prime source of re-

[1] Mr. Cornford's book might in fact be entitled with greater propriety *From Philosophy to Religion.*

ligious and magical ideas, but it does not follow that the idea of God is arrived at by way of a notion of collective *mana*. No doubt that would be the probable course of events, if the savage had as little sense of his individuality as Cornford supposes; but it seems to me rather that the savage's strong sense of individuality has led at an early stage to the personalisation, the individuation, of *mana*, the vaguely conceived spiritual power and influence, and that it was only by a long course of religious and philosophical speculation that men reached the conception of the Absolute or of God as a universal power of which each personal consciousness is a partial manifestation.

It is interesting to note that, if we could accept Cornford's views, we could now claim to witness the completion of one full cycle of the wheel of speculation, the last step having been made in an article in a recent number of the *Hibbert Journal*;[1] for it is there suggested that the only God or super-individual power we ought to recognise and revere is, not a collective consciousness conceived as a supra-individual unity of consciousness, but the collective mind of humanity in the sense in which I am using the term, a system of mental forces that slowly progresses towards greater harmony and integration.

M. Lévy Bruhl has written an interesting, though highly speculative, account of savage mental life which he represents as differing profoundly from our own, chiefly in that it is dominated by "collective representation."[2] His view is not unlike that put forward by Cornford.

Collective representations or ideas are rightly said to be the product of the group mind rather than of any individual mind; that is to say, they have been gradually evolved by collective mental life; and they are said to differ from our ideas in being "states more complex in

[1] Oct., 1914.

[2] *Les Fonctiones mentales dans les Sociétés inférieures*, Alcan, Paris, 1910.

which the emotional and motor elements are integral parts of the ideas." Thinking by aid of these collective representations is said to have its own laws quite distinct from the laws of logic.

These statements are no doubt correct; but both Lévy Bruhl and Cornford commit the great error of assuming that the mental life of civilised man is conducted by each individual in a purely rational and logical manner; they overlook the fact that we also are largely dominated by collective representations; for these collective representations are nothing but ideas of objects to which traditional sentiments, sentiments of awe, of fear, of respect, of love, of reverence, are attached. Almost the whole of the religion and morality of the average civilised man is based on his acquisition of such collective representations, traditional sentiments grown up about ideas of objects, ideas which he receives ready made and sentiments which are impressed upon him by the community that has evolved them.

It is no doubt true that in the main the field of objects to which collective representations apply is larger in savage life; and these ideas are more uniform and more powerful and unquestioned, because the group is more homogeneous in its sentiments. But it is, fortunately, only a rare individual here and there among us who in considerable degree emancipates himself from the influence of such representations and becomes capable of confronting all objects about him in a perfectly cool, critical, logical attitude—who can "peep and botanise upon his mother's grave." Only by strict intellectual discipline do we progress towards strictly logical operations in relation to real life, towards pure judgments of fact as opposed to judgments of value. For our judgments of value are rooted in our sentiments; and whatever is for us an object of a sentiment of love or hate, of attachment or aversion, can only with the greatest difficulty, if at all, be made an object of a pure judgment of fact.

And Cornford and Lévy Bruhl make the same mistake in regard to "collective representations" as in regard to the group self-consciousness—namely, they credit the savage with theories for the explanation of the beliefs implicitly involved in the "collective representations," for example, the theory of mystic participation, which is said to replace for the savage the civilised man's theory of mechanical causation. But, when we regard any material object as holy, or sacred, or as of peculiar value, because it was given us by a departed friend, or belonged to and perhaps was once worn by a beloved person, our behaviour towards it is not determined by any theory of participation; if, for example, we touch it tenderly and with reverent care, that is the direct expression of our feeling. We even behave as if we held the theory of participation, to the extent of believing that the dead or distant person will suffer pain if we ill-use or neglect the object which is associated with him in our minds, but without actually holding that belief; and still more without elaborating a theory of the nature of the process by which our action will produce such an effect. It is only a late and highly sophisticated reflection upon behaviour of this kind which leads to theories for the justification of such behaviour. It is not true, then, that we are logical individuals, while savages are wholly prelogical in virtue of the dominance among them of the collective mental life. The truth rather is that, wherever emotion qualifies our intellectual operations, it renders them other than purely and strictly logical; and the savage or the civilised man departs more widely from the strictly logical conduct of his intellect, in porportion as his conceptions of things are absorbed without critical reflection and analysis and are coloured with the traditional sentiments of his community. The average savage, being more deeply immersed in his group, suffers these effects more strongly than the average civilised man. Yet the interval in this respect between the modern man of scientific culture and the average citizen of our modern states is far greater than that between the latter and the savage.

If one had to name the principal difference between the conditions of life of the typical savage and those of the average civilised man, one would, I think, have to point to the lack in civilised life of those conditions which so inevitably develop the group-consciousness of the savage. The family circle supplies to the young child something of these conditions, but in a very imperfect degree only. At an early age this influence is much weakened by general intercourse. As the individual approaches maturity, he finds himself at liberty to cut himself off completely from all his natural setting, to transplant himself to any part of the world, and to share in the life of any civilised community. He can earn his livelihood anywhere, and he knows nothing of communal responsibility.

Progressive weakening of the conditions that force the development of group self-consciousness has characterised the whole course of the development of civilisation, and has reached its climax in the conditions of life in our large cities.

In primitive communities the conditions of group self-consciousness are, as we have seen, fourfold; namely kinship, territorial, traditional and occupational association. All these are present in the highest degree in the nomadic group under the typical patriarchal system.

When kinship groups take to agriculture and become permanently settled on one spot, the kinship factor tends to be weakened, through the inclusion of alien elements; and the territorial factor becomes the most important condition. Throughout European history the territorial factor, expressing itself in the form of the village-community, remained of universal importance in this respect; the Roman Empire and the Roman Church weakened it greatly; but everywhere outside those spheres it continued to be of dominant importance until the great social revolution of the modern industrial period.

The village-community maintained much of the tradition and custom that tend to develop group self-consciousness with its moralising influence. But at the present time almost the only condition of wide and general influence that continues in times of peace to foster group self-consciousness is occupational association. And so we find men tending more and more to be grouped for all serious collective activities according to their occupations. From the earliest development of European industry this tendency has been strong; it produced the trading and craft guilds which played so great a part in medieval Europe; and, though the monarchical and capitalistic régimes of modern times have done all they can to repress and break up these occupational groups, and have greatly restricted their influence, they have failed to suppress them entirely. The climax of this tendency for the occupational to replace and overshadow all other forms of self-conscious grouping is present-day Syndicalism.

The natural conditions of group self-consciousness, which in primitive societies rendered its development inevitable and spontaneous in every man, have then been in the main destroyed. But man cannot stand alone; men cannot live happily as mere individuals; they desire and crave and seek membership in a group, in whose collective opinions and emotions and self-consciousness and activities they may share, with which they may identify themselves, thereby lessening the burden of individual responsibility, judgment, decision, and effort.

Hence in this age the natural groupings and the involuntary developments of group-consciousness are largely replaced by an enormous development of artificial voluntary groupings, over and above the natural groupings that are still only in very imperfect measure determined by the weakened force of the natural conditions, namely kinship, neighbourhood, and occupation.

In part these artificial groupings are designed to reinforce the natural conditions, as, for example, village festivals. The whole population of a country such as our own is permeated by a vast and complexly interwoven, or rather tangled, skein of the bonds of voluntary associations. Many of these are, of course, formed to undertake some definite work, to achieve some end which can only be achieved by co-operative effort. But in the majority of such cases the satisfactions yielded by group life play a very great part in leading to the formation of, and in maintaining, the groups; for example, groups of philanthropic workers, the makers of charity bazaars, the salvation army, the churches, the chapels, the sects. Most of such associations that have any success and continuity of existence contain a nucleus of persons who identify themselves in the fullest possible manner with the group, make its interest their leading concern, the desire of its welfare their dominant motive, and find in its service their principal satisfaction and happiness.

And in very many voluntary associations the group motives, the desire for the satisfactions to be found in group life, are of prime importance, predominating vastly over the desire to achieve any particular end by co-operative action. Such are our countless clubs and societies formed frankly for recreation, or for mutual improvement, and for all kinds of ostensible purposes which serve merely as excuses or reasons for the existence of the club. In the majority of instances these declared purposes really serve merely or chiefly to exert a certain natural selection of persons, to bring together persons of similar tastes as voluntary associates, to enable, in short, birds of a feather to flock together. Even some of our enduring historical institutions owe their continuance chiefly to the advantages and satisfactions that proceed from group-consciousness, for example, colleges, school-houses, and political parties, especially perhaps in America. Party feeling, as Sir H. Maine rightly said, is frequently a remedy for the inertia of democracy.

The savage, when he maintains associations other than those determined by natural conditions, intensifies his group-consciousness by wearing badges and totem marks, by tattooing and scarring, and by indulging in various rites and ceremonies, about which a certain secrecy and mystery is maintained. And civilised men exhibit just the same tendency and take very similar measures to intensify group-consciousness. We have our club colours and ribbons and blazers, our college gowns and colours, our Oxford accent, our badges of membership, and so on. Freemasonry, with its lodges and badges and mysterious rites, seems to be the purest example on a large scale. And, when the group-consciousness and the group sentiment have been acquired, we continue to cultivate it purely for its own sake, by holding annual dinners and reunions of old boys, and so forth.

It is of the greatest importance that this tendency to seek and maintain a share in group-consciousness, which, as we have seen, manifests itself everywhere even under the most adverse conditions, not merely yields comfort and satisfaction to individuals, but brings about results which are in almost every way extremely advantageous for the higher development of human life in general.

We have seen that, in the well-organised group, collective deliberation, judgment, and action are raised to a higher plane of effectiveness than is possible to the average member of the group. But apart from that, the group spirit continues with us, as with the savage (though in a less effective degree) to be the great socialising agency. In the majority of cases it is the principal, if not the sole, factor which raises a man's conduct above the plane of pure egoism, leads him to think and care and work for others as well as for himself. Try to imagine any man wholly deprived of his group-consciousness and set over against all his fellow-men as an individual unit, and you will see that you could expect but little from him in the way of self-sacrifice or public service—at most a care for his wife and children and sporadic acts of kindliness when direct appeals are made to his pity; but none of that energetic and devoted public service and faithful self-sacrificing co-operation without which the continued welfare of any human society is impossible.

The group spirit destroys the opposition and the conflict between the crudely individualistic and the primitive altruistic tendencies of our nature.

This is the peculiar merit and efficiency of the complex motives that arise from the group spirit; they bring the egoistic self-seeking impulses into the service of society and harmonise them with the altruistic tendency. The group spirit secures that the egoistic and the altruistic tendencies of each man's nature, instead of being in perpetual conflict, as they must be in its absence, shall harmoniously co-operate and reinforce one another throughout a large part of the total field of human activity.

For it is of the essence of the group spirit that the individual identifies himself, as we say, with the group more or less; that is to say, in technical language, his self-regarding sentiment becomes extended to the group more or less completely, so that he is moved to desire and to work for its welfare, its success, its honour and glory, by the same motives which prompt him to desire and to work for his own welfare and success and honour; as in the case of the student working for a scholarship or university prize, or the member of an exploring expedition or fighting group. Further, the motives supplied by the group spirit may be stronger than, and may overpower, the purely individualistic egoistic motives, just because they harmonise with, and are supported by, any altruistic tendency or tendencies comprised in the make-up of the individual; which altruistic tendencies will, where the group spirit is lacking, oppose and weaken the effects of purely egoistic motives. To illustrate this principle, let us imagine an Englishman who, in a Congo forest, finds a white man sick or in difficulties. To succour the sick man may be to incur grave risks, and he is tempted to pass on; but the thought comes to him that in so doing he will lower the prestige of the white man in the eyes of the natives; and this idea, evoking the motives of the group spirit which unites all white men in such a land, brings victory to his sense of pity in its struggle with selfish fear.

In this way, that is by extension to the group, the egoistic impulses are transmuted, sublimated, and deprived of their individualistic selfish character and effects and are turned to public service. Hence it is that it is generally so difficult or impossible to analyse the motives of any public service or social activity and to display them as either purely egoistic or altruistic; for they are, as Herbert Spencer called them, ego-altruistic. And hence it comes about that both the cynic and the idealist can make out plausible cases, when they seek to show that either egoism or altruism predominates in human life. Both are right in a partial sense.

Another noteworthy feature of the group spirit renders it extremely effective in promoting social life; namely the fact that, although the group sentiment is apt to determine an attitude of rivalry, competition, and antagonism towards similarly constituted groups, yet a man may share in the self-consciousness of more groups than one, so long as their natures and aims do not necessarily bring them into rivalry. And in our complex modern societies this principle of multiple group-consciousness in each man is of extreme importance; for without it, and in the absence or comparative lack of the natural conditions of grouping other than the occupational, the whole population would become divided into occupational groups, each fighting collectively against every other for the largest possible share of the good things of life. A tendency towards this state of things is very perceptible, in spite of the correcting cross-connections of kinship, of church and political party, and of territorial association.

But another principle of multiple group-consciousness is, perhaps, of still greater importance, namely that it allows the formation of a hierarchy of group sentiments for a system of groups in which each larger group includes the lesser; each group being made the object of the extended self-regarding sentiment in a way which includes the sentiment for the lesser group in the sentiment for the larger group in which it is comprised. Thus the family, the village, the county, the country as a whole, form for the normal man the objects of a harmonious hierarchy of sentiments of this sort, each of which strengthens rather than weakens the others, and yields motives for action which on the whole co-operate and harmonise rather than conflict.

Such a hierarchy is seen in savage life. It often happens that a man is called on to join in the defence of some village of the tribe other than his own. In such cases he is moved not only by his tribal sentiment, but also by his sentiments for his village and family. The sentiment for the part supports the sentiment for the whole.

It is of considerable importance also that in general the

development of a sentiment of attachment to one group not only does not prevent, but rather facilitates, the development of similar sentiments for other groups. And this is especially true when the groups concerned are related to one another as parts and wholes, that is, when they form a hierarchy of successively more widely inclusive groups. The sentiment for the smaller group (e. g. the family) naturally develops first in the child's mind; if only for the reason that this is the group of which he can first form a definite idea, and with the whole of which he is in immediate relations. The strong development of this first group sentiment prepares the child's mind for the development of other and wider group sentiments. For it increases his power of grasping intellectually the group of persons as a complex whole; and it strengthens by exercise those impulses or primary tendencies which must enter into the constitution of any group sentiment; and, thirdly, it prevents the excessive development of the purely individualistic attitude, of the habit of looking at every situation and weighing all values from the strictly individualistic and egoistic standpoint; which attitude, if once it becomes habitual, must form a powerful hindrance to the development of the wider group sentiments, when the child arrives at an age to grasp the idea of the larger group.

The organisation of an army again illustrates these principles in relatively clear and simple fashion. In our own army the regiment is the traditional self-conscious unit about which traditional sentiment and ritual have been carefully fostered, in part through realisation of their practical importance, in part because this unit is of such a size and nature as to be well suited to call out strongly the natural group tendencies of its component individuals. On the whole the military authorities, and especially Lord Haldane in the formation of the territorial army, seem to have wisely recognised the importance of the group spirit of the regiment; although during the Great War it was, under the pressure of other considerations, apparently lost sight of at certain times and places, with, I believe, deplorable consequences.

In modern warfare, and especially in the Great War, the Division has tended to become of predominant importance as the unit of organisation; and accordingly, without destroying or superseding regimental group-consciousness, the sentiment for the Division has been in many instances a very strong factor in promoting the spiritual cohesion and efficiency of the army. Certain Divisions, such as the 10th and the 29th, have covered themselves with glory, so that the soldiers have learnt to feel a great pride in and a devotion to the Division.

This larger group, although of comparatively ephemeral existence and therefore devoid of long traditions coming down from the past, is in perfect and obvious harmony, in purpose and spirit and material organisation, with the battalions and other units of which it is composed; and, accordingly, the sentiment for the larger group does not enter into rivalry with that for the battalion, the battery, or other smaller unit; rather it comprises this within its own organisation and derives energy and stability from it.

These psychological principles of group-consciousness are, I think, well borne out inductively, by any comparative survey; that is to say, we find that, where family sentiment and the sentiment for the local group are strong, there also the wider group sentiments are strong, and good citizenship, patriotism, and ready devotion to public services of all kinds are the rule. The strongest most stable States have always been those in which family sentiment has been strong, especially those in which it has been strengthened and supported by the custom of ancestor worship, as in Rome, Japan and China. Scotchmen again (Highlanders especially) are noted for clannishness, and Scotch cousins have become a byword; a fact which implies the great strength of the family sentiment. The clan sentiment, which is clearly only an extension of the family sentiment, is also notoriously strong. The sentiment for Scotland as a whole is no less strong in the hearts of all her sons. But, and this is the important point, these strong group sentiments are perfectly compatible with and probably conducive to a sentiment for the still wider group, Great Britain or the British Empire; and the public services rendered to these larger groups by men of Scottish birth are equally notorious.

In these considerations we may see, I think, a principal ground of the importance of the institution of the family for the welfare of the state. The importance has often been insisted upon; but too much stress is usually laid upon the material aspects, and not enough upon the mental effects, of family life.

It has been a grave mistake on the part of many collectivists, from Plato onwards, that they have sought to destroy the family and to bring up all children as the children of the state only, in some kind of barrack system. It is not too much to say that, if they could succeed in this (and in this country great strides in this direction are being rapidly made), they would destroy the mental foundations of all possibility of collective life of the higher type.

We touch here upon a question of policy of the highest importance. There are, it seems to me, three distinct policies which may be deliberately pursued, for the securing of the predominance of public or social motives over egoistic motives. First, we may aim at building up group life on the foundation of a system of discipline which will result in more or less complete suppression of the egoistic tendencies of individuals, the building up in them of habits of unquestioning obedience to authority. I imagine that the Jesuit system of education might fairly be taken as the most successful and thorough-going application of this principle. The organisation of an army of unwilling conscripts to fight for a foreign power must rest on the same basis. Some group spirit no doubt will generally grow up. But, though wonderful results have been obtained in this way, the system has two great weaknesses. First, it seeks to repress and destroy more than half of the powerful forces that move men to action—namely, the egoistic motives in general—instead of making use of them, directing them to social ends. Secondly, it necessarily crushes individuality and therefore all capacity of progress and further development in various directions; it results in a rigidly conservative system without possibility of spontaneous development.

The second system is that which aims at developing in all members of the state or inclusive group a sentiment of devotion to the whole, while suppressing the growth of sentiments for any minor groups within the whole. This was the system of Plato's *Republic* and is essentially the collectivist ideal. It is the policy of those who would suppress all sentimental groupings, all local loyalties and patriotisms, in favour of the ideal of the brotherhood of man, the cosmopolitan ideal. I have already pointed out one great weakness of this plan—namely, that this sentiment for the all-inclusive group cannot be effectively developed save by way of development of the minor group sentiments. And, though it may succeed with some persons, there will always be many who cannot grasp the idea of the larger whole sufficiently firmly and intelligently to make it the object of any strong and enlightened sentiment of attachment; such persons will be left on the purely egoistic level, whereas their energies might have been effectively socialised by the development of some less inclusive group-consciousness.

Again, the smaller group is apt to call out a man's energies more effectively, because he can see and foresee more clearly the effects of his own actions on its behalf. Whereas the larger the group, the more are the efforts of individuals and their effects obscured and lost to view in vast movements of the collective life. That is to say, the smaller groups harmonise more effectively than the larger groups the purely egoistic and the altruistic motives (except of course in the case of those few persons who can play leading parts in the life of the larger group). For, though a man may be moved by his devotion to the group to work for its welfare, he will work still more energetically if, at the same time, he is able to achieve personal distinction and acknowledgment, if the purely egoistic motives can also find satisfaction in his activities. Hence this second policy also, no matter how successful, fails to make the most of men, fails to bring to the fullest exercise all their powers in a manner that will promote the welfare of the whole. Thirdly, this system loses the advantages of the healthy rivalry between groups within the whole; which rivalry is a means to a great liberation of human energies. These are the weaknesses of the over-centralised state, such as modern France or the Roman Empire.

Only the third policy can liberate and harmonise the energies of men to the fullest extent; namely, that which aims at developing in each individual a hierarchy of group sentiments in accordance with the natural course of development.

One other virtue of the group spirit must be mentioned. Although it tends to bring similar groups into keen rivalry and even into violent conflict, the antagonism between men who are moved to conflict by the group spirit is less bitter than that between individuals who are brought into conflict by personal motives; for the members of each group or party, though they may wish to frustrate or even to destroy the other party as such, may remain benevolent towards its members individually. And this is rendered easier by the fact that the members of each group, recognising that their antagonists are also moved by the group spirit, by loyalty and devotion to the group, will sympathise with and respect their motives far more readily and fully than they would, if they ascribed to their opponents purely egoistic motives. This recognition, even though it be not clearly formulated, softens the conflict and moderates the hostile feelings that opposition inevitably arouses in men keenly pursuing any end, especially one which they hold to be a public good; in this way it renders possible that continuance of friendly relations between members of bitterly opposed parties which has happily been the rule and at the same time the seeming anomaly of English public life.

In our older educational system, and especially in the "public schools" and older universities, the advantages and the importance of developing the group spirit have long been practically recognised—*esprit de corps* has been cultivated by the party system, by rivalry of groups within the group; by forms, schoolhouses, colleges, clubs, teams, games, and by keeping the honour and glory of the school, college, or other unit, prominently before the minds of the scholars in many effective ways. It is, I think, one of the gravest defects of our primary system of education that it makes so little provision for development of this kind; that, while it weakens the family sentiment, it provides no effective substitute for it. Something has been done in recent years to remedy this defect, notably the fostering of the boy-scout movement; but every opportunity of supplying this need should be seized by those who are responsible for the direction of educational policy.

The importance of the group spirit may be illustrated by pointing to those individuals and classes which are denied its benefits. The tramp, the cosmopolitan globe-trotter, the outcast in general, whether the detachment from group life be due to the disposition or choice of the individual or to unfortunate circumstances, is apt to show, only too clearly, how little man is able, standing alone, to maintain a decent level of conduct and character. On a large scale this is illustrated by the casteless classes in caste communities, and especially by Eurasians of India and by other persons and classes of mixed descent, who fail to identify themselves wholly with either of the groups from which they derive their blood. The moral defects of persons of these classes have often been deplored, and they have usually been attributed to the mixture of widely different hereditary strains. There is probably some truth in the view; but in general the moral shortcomings of persons of these classes are chiefly due to the fact that they do not fully share in the life of any group having old established moral traditions and sentiments.

Summary of Principal Conditions of the Development of the Group Spirit

We have seen that the group spirit plays a vastly important part in raising men above the purely animal level of conduct, in extending each man's interests beyond the narrow circle of his own home and family, in inspiring him to efforts for the common good, in stimulating him to postpone his private to public ends, in enabling the common man to rise at times, as shown by a multitude of instances during the Great War, to lofty heights of devotion and self-sacrifice.

The development of the group spirit consists in two essential processes, namely, the acquisition of knowledge of the group and the formation of some sentiment of attachment to the group as such. It is essential that the group shall be apprehended or conceived as such by its members. Therefore the group spirit is favoured by whatever tends to define the group, to mark it off distinctly from other groups; by geographical boundaries; by peculiarities of skin-colour or of physical type, of language, or accent, of dress, custom, or habit, common to the members of the group; that is to say, by homogeneity and distinctiveness of type within the group.

And, though definition of the group as such within the minds of its members is the prime condition of the growth of the group spirit, that spirit will be the more effective the fuller and truer is the knowledge of the group in the minds of its members. Just as individual self-knowledge favours self-direction and wisdom of choice, so group self-knowledge must, if it is to be fully effective, comprise not only the conception of the group as a whole but also the fullest possible knowledge of the component parts and individuals and an understanding of their relations to one another. In this respect smallness and homogeneity of the group are obviously favourable. But knowledge of the group, however exact and widely diffused, is of itself of no effect, if there be not also widely diffused in the members some sentiment of attachment to the group. The prevailing group sentiment is almost inevitably one of attachment. There are exceptional instances in which men are compelled to act as members of a group which they hate or despise, notably in some cases of compulsory military service and in convict gangs; but it must be rare that, even under such conditions, some sense of common interest, some fellow feeling for other members in like distressing circumstances, does not lead to the growth of some group spirit, provided only that the group has some continuity and some homogeneity in essentials.

In all natural and spontaneously formed groups, the extension of the self-regarding sentiment to the group is a normal and inevitable process; and, like the self-regarding sentiment, the sentiment so formed may range from an insane and incorrigible pride (as often in the case of the family sentiment) to a decent self-respect that is perfectly compatible with a modest attitude and with reasonable claims upon and regard for the interests of other groups.

The main difference between the self-regarding sentiment and the developed group sentiment is that the latter commonly involves an element of devotion to the group for its own sake and the sake of one's fellow members. That is to say, the group sentiment is a synthesis of the self-regarding and the altruistic tendencies in which they are harmonised to mutual support and re-enforcement: the powerful egoistic impulses being sublimated to higher ends than the promotion of the self's welfare.[1]

Further, the group has, or may have, a greater continuity of existence than the individual, both in the past and in the future; and for this reason, and because also it includes the purely altruistic tendency, the group sentiment is capable of idealisation in a high degree and of yielding satisfactions far more enduring and profound than the most refined self-sentiment.

Both knowledge of the group and the growth of the group sentiment are greatly promoted by two processes, in the absence of which the group spirit can attain only a very modest development. These are free intercourse within the group and free intercourse between the group and other groups. We shall have occasion to discuss and illustrate them in later chapters.

[1] At this point I would refer the reader to the discussion of the self-regarding sentiment (Chapter VII) in my *Social Psychology*.

CHAPTER V

Peculiarities of Groups of Various Types

WE have discussed the psychology of the simple crowd or unorganised group; and taking an army as an extreme and relatively simple type of the highly organised group, we have used it to illustrate the principal ways in which organisation of the group modifies its collective life, raising it in many respects high above that of the crowd.

I propose now to discuss very briefly the peculiarities of groups of several types. Some classification of groups seems desirable as an aid to the discovery of the general principles of collective life and their application to the understanding of social life in general. It seems impossible to discover any single principle of classification. Almost every group that enjoys a greater continuity of existence than the simple crowd partakes in some degree of qualities common to all. But we may distinguish the most important qualities and roughly classify groups according to the degrees in which they exhibit them.

Apart from crowds, which, as we have seen, may be either fortuitously gathered or brought together by some common purpose, there are many simple groups which, though accidental in origin (i.e. not brought together by common purpose or interest) and remaining unorganised, yet present in simple and rudimentary form some of the features of group life.

The persons seated in one compartment of a railway

train during a long journey may be entirely strangers to one another at the outset; yet, even in the absence of conversation, they in the course of some hours will begin to manifest some of the peculiarities of the psychological group. To some extent they will have come to a mutual understanding and adjustment; and, when a stranger adds himself to their company, his entrance is felt to some extent as an intrusion which at the least demands readjustments; he is regarded with curious and to some extent hostile glances. If an outsider threatens to encroach on the rights of one of the company, the others will readily combine in defence of their member; and any little incident affecting their one common interest (namely, punctual arrival of the train at its destination) quickly reveals, and in doing so strengthens, the bond of common feeling.

On a sea voyage the group spirit of the passenger ship attains a greater development, by reason of the longer continuance of the group, its more complete detachment and definition, the sense of greater hazard affecting all alike, the sense of dependence on mutual courtesy and good-will and sympathy for the comfort and enjoyment of all. Very soon the experienced traveller, contrasting and comparing this present company with those of previous voyages, sums up its qualities and defects and lays his plans accordingly. And by the time that an intermediate port is reached, where perhaps the most "grumpy" and least entertaining member of the company disembarks, even his departure is felt by the rest as a loss that leaves a gap in the structure of the group.

Such fortuitous and ephemeral groups apart, all others may be classed in the two great divisions of natural and artificial groups.

The natural groups again fall into two main classes which partly coincide—namely, those rooted in kinship and those determined by geographical conditions. The family is the pure example of the former; the population of a small island, the type of the latter kind. The main difference is that the bonds of the kinship group are purely or predominantly mental and therefore can, and commonly do, remain effective in spite of all spatial separation and of all lack of common purpose or of material benefits accruing from membership in the group.

The artificial groups may be divided into three great classes, the purposive, the customary or traditional, and the mixed; those of the last kind combining the purposive and the traditional characters in various proportions.

The purposive group is brought together and maintained by the existence of a common purpose in the minds of all its members. It is, in respect of efficiency, the highest type; for it is essentially self-conscious, aware of its ends and of its own nature, and it deliberately adopts an organisation suited to the attainment of those ends. The simplest and purest type is the social club, a body of people who meet together to satisfy the promptings of the gregarious instinct and to enjoy the pleasures of group life. In the great majority of instances, the social club adopts some form of recreation—debating, music, chess, whist, football, tennis, cycling—the practice of which gives point and definition to the activities of members and secures secondary advantages. It is noteworthy that on this purely recreational plane, clubs and societies of all sorts seek in almost all cases to enhance the group-consciousness and hence the satisfactions of group life by entering into relations, generally relations of friendly rivalry, but sometimes merely of affiliation and formal intercourse, with other like groups. For not only is the group-consciousness enriched and strengthened by such intercourse; but, when the rival or communicating groups, becoming aware of one another, become informally or, more generally, formally allied to constitute a larger whole, the consciousness of participation in this larger whole gratifies more fully the gregarious impulse and enhances the sense of power and confidence in each member of each constituent group. This seems to be the main ground of that universal tendency to the formation of ever more inclusive associations of clubs and societies, which, overleaping even national boundaries and geographical and racial divisions, has produced numerous world-wide associations.

Another very numerous class of strictly purposive groups is to be found in the commercial companies. In these the group spirit commonly remains at the lowest level; for the dominant motive is individual financial gain, and the only common bond among the shareholders is their interest in the management of the company so far as it affects the private and individual end of each one. Group self-knowledge, organisation, tradition, and group sentiment are all at a minimum; accordingly the group remains incapable of effective deliberation or action. It operates through its board of directors and officers and, owing to its incapacity for group action, has to rely upon the provisions of the Company Laws for the control of their actions.

A third large class of purposive groups are the associations formed for the furthering of some public end. Many such groups are purely altruistic or philanthropic; but in the majority the members hope to share in some degree in the public benefits for the attainment of which the group is formed. In many such associations, group life hardly rises above the low level of the commercial company; the main difference being that, in virtue of the "disinterested" or public-spirited nature of the dominant purpose, the members regard one another and their executive officers with greater confidence and sympathy; even though remaining personally unacquainted. Notable instances of such associations, achieving great public ends, are "The

National Trust for the Preservation of Places of Natural Beauty or Historic Interest," and "The Public Footpaths Association." Other associations of this kind have something of the nature of a commercial company: e.g. "The First Garden City Company," and "The Trust Houses Company." The peculiarity of these is that the motive of financial gain is subordinated to, while co-operating with, the desire for achievement of a public good, a benefit to the whole community in which the members of the group share in an almost inappreciable degree only. Such associations are very characteristic of the life of this country; and it may be hoped that their multiplication and development will prove to be one of the ameliorating factors of the future, softening the asperities of commercial life, correcting to some degree that narrowing of the sympathies, and preventing that tendency to class antagonisms, which purely commercial associations inevitably produce. The great co-operative societies seem to have something of this character; for, although the dominant motive of membership is probably in most cases private advantage, yet membership brings with it some sense of participation in a great movement for better social organisation, some sense of loyalty to the group, some rudimentary group knowledge and group spirit, some interest in and satisfaction in the prosperity of the group for its own sake, over and above the strictly private interest of each member. The introduction of various forms of profit sharing will give something of this character to commercial companies.

The recent investments in government loans by millions of individuals, acting in part from patriotic motives, must have a similar tendency; and a similar effect on a large scale must be produced by any nationalisation of industries, a fact which is one of the weightiest grounds for desiring such nationalisation; though it remains uncertain whether, when the scale of the association becomes so large as to include the whole nation, the bulk of the citizens will be able effectively to discern the identity of their public and private interests, and whether, therefore, such nationalisation will greatly promote that fusion or co-operation of public and private motives which is the essential function and merit of the group spirit.

The most characteristic British group of the purposive type is the association formed for some public or quasi-public end and operating through a democratically elected committee or committees and sub-committees. Such groups are the cradle of the representative principle and the training ground of the democratic spirit, especially of its deliberative and executive faculties. In them each member, taking part in the election of the committee, delegates to them his share of authority, but continues to exert control over them by his vote upon reports of the committee and in the periodic re-election of its members. On this ground the citizens are trained to understand the working of the representative principle; to yield to the opinion of the majority on the choice of means, without ceasing loyally to co-operate towards the common end; to observe the necessary rules of procedure; to abide by group decisions; to influence group opinion in debate, and in turn to be influenced by it and respect it; to differ without enmity; to keep the common end in view, in spite of the inevitable working of private and personal motives; to understand the necessity for delegation, and to respect the organisation through which alone the group raises itself above the level of the crowd.

Traditional groups of pure or nearly pure character are relatively infrequent. Perhaps the castes of the Hindu world are, of all large groups, those which most nearly approach the pure type. Traditional grouping is characteristic of stagnant old established populations, of which it is the basis of organisation and principal cement. No doubt in almost every case the formation of the traditional group was in some degree purposive; but the original purpose has generally been lost sight of; myths and legends have grown up to explain the origin of and give a fictitious purpose or *raison d'être* to the group. In the absence of any definite practical purpose animating the group and holding it together, its stability is secured and its tradition is re-enforced and given a visible presentation by the development of ritual. Of all the great groups among us the Free Masons perhaps afford the best illustration of this type.

Far more important in the British world are the groups of the mixed type, partly traditional and partly purposive, groups having a long history and origins shrouded in the mists of antiquity, but having some strong and more or less definite common purpose. Of such groups the Christian Church is the greatest example. In the Roman Church, whose history has been so little interrupted, tradition attains its fullest power, and the regard for the past is strengthened and supplemented by the prospect of an indefinitely prolonged future directed towards the same ends. Its organisation has grown gradually under the one continued overshadowing purpose, every addition becoming embodied and established in the great tradition, the strength of which is perpetually maintained by ritual. And this traditional organisation is not only borne in the minds of each generation of members of the Church, but, in an ever increasing degree, has embodied itself in a material system of stone and glass and metal and printed words; these constitute a visible and enduring presentment which, though entirely disconnected and heterogeneous in a merely material sense, yet provides fixed points in the whole organisation, contributing immensely to its stability, and aiding greatly in bringing home to the minds of its members the unity of the whole group in the past, the present, and the future. Many groups or sects having the same essential purpose as the Roman Church have aimed to establish a tradition without the aid of such material embodiments; but their ephemeral histories illustrate the wisdom of the mother Church which, in building up her vast organisation, has recognised the limitations and the frailties of the human mind and has not scorned to adapt herself to them in order to overcome them.

On a smaller scale our ancient universities and their colleges illustrate the same great type of the partly traditional partly purposive group, and the same great principles of collective life—namely, the stability derived from the continuity of tradition, from its careful culture, and its partial embodiment in ritual and material structure.

An essential weakness of all such groups in a progressive community is that tradition tends to overshadow purpose; hence every such group tends towards the rigidity and relative futility of the purely traditional group. Its organisation tends to set so rigidly that it is incapable of adapting itself to the changing needs of the present and the future; the maintenance of tradition, which is but a means towards the acknowledged end, becomes an end in itself to which the primary purpose of the whole is in danger of being subordinated.

The churches and the universities alike illustrate vividly the principles of a group within a group. Each of the older universities is a microcosm, a small model of the national life; and largely to this fact is due its educational value as a place of residence. Each college evokes a strong group spirit in all its members; and this sentiment for the college, though it may and does in some minor matters conflict with the sentiment for the university, is in the main synthesised within this, and indeed is the chief factor in the strength of that sentiment.

The group spirit of each college owes much of its strength to the carefully fostered, but perfectly friendly, rivalry between the several colleges in sports and studies and other activities. The close companionship and emulation between a number of small communities of similar constitution and purpose, each having a long and distinct tradition as well as a clearly defined material habitat which embodies and symbolises its traditions in a thousand different ways, has raised the self-knowledge and sentiment of the groups to a high level. It is well known that the few years spent in one of the colleges develops in every member (with few exceptions) a sentiment of attachment that persists through life and extends itself in some degree to every other member, past, present, and future; so that, in whatever part of the world and under whatever circumstances two such men may meet, the discovery of their common membership of the college at once throws them into a friendly attitude towards each other and prepares each to make disinterested efforts on behalf of the other.

The same is true in a less degree of the universities themselves. Oxford and Cambridge have, partly in consequence of their proximity and close intercourse, developed on closely parallel lines. They are therefore so similar in constitution and aims as to be keen though friendly rivals. This has been of great benefit to both, the self-knowledge and group sentiment of each having been greatly promoted by this close intercourse, rivalry, and reciprocal criticism. And this rivalry has not prevented the growth of some sentiment for the larger group constituted by the members of both universities, each of whom is always ready to defend the common interests of the larger group against the rest of the world.

Again, within each college there are numerous smaller groups, each with its traditions and group spirit; and, so long as these groups do not become too exclusive, do not absorb all the devotion of their members, but leave each one free to join in the life of other minor groups, their influence is good, the group spirit of each such minor group contributing to the strength of the larger group sentiment and enriching the spiritual life of the whole.

In the middle ages occupational groups were of great importance and influence. They were of the mixed type, for most of them, though essentially purposive, developed strong traditions; and in their remote origins many of them were perhaps rather natural than artificial formations. The violent changes of industrial life, the development of the capitalist system and modern industrialism, dislocated and largely destroyed these occupational groups to the great detriment of social well-being. At the present time we see a strong tendency to the growth of occupational groups of the purely purposive type, which, lacking the guidance and conservative power of old traditions, and depending for their strength largely upon the identification of the material interests of each member with those of the group, show a narrowness of outlook, a lack of stability and internal cohesion, and a tendency to ignore the place and function of the group in the whole community. They show, in short, a lack of the enlightened group spirit which only time, with increasing experience and understanding of the nature and functions of group life, can remedy. It may be hoped that with improved internal organisation, with the growth of more insight into the mutual dependence of the various groups on one another and on the whole community, these groups, which at present seem to some observers to threaten to destroy our society and to replace the rivalry of nations by an even more dangerous rivalry of vast occupational groups, may become organised within the structure of the whole and play a part of the greatest value in the national life.

PART II

THE NATIONAL MIND AND CHARACTER

CHAPTER VI

Introductory

What is a Nation?

HAVING studied the most general principles of the collective mental life, as exemplified in the two extreme forms of the unorganised crowd and the highly organised army, having briefly noted the principal classes of groups that enjoy a collective mental life, and having examined the nature and function of the group spirit in the organisation of the group mind, we may now take up the study of the most interesting, most complex, and most important kind of group mind, namely the mind of a nation-state.[1]

Many attempts have been made to define more exactly the popular notion of a nation. The word has sometimes been applied to large groups of primitive folk that show evidence of close racial affinity and similarity of customs, such as the Iroquois tribes of North America, or the Hun invaders of medieval Europe. In popular usage the word is more commonly restricted to the great nation-states of modern times. It must be recognised that, since human societies of present and past times present every conceivable variety of composition and structure, it is impracti-

[1] For a brief history of the nation-state the reader may be referred to Prof. Ramsay Muir's *Nationalism and Internationalism*, London, 1917. He rightly describes "nationalism" as one of the most powerful factors in modern history. It is, I think, obviously true that we may go further and say that it is *the* most powerful factor in modern history.

cable to lay down any strict definition and to classify populations as falling definitely within or outside the class.

But, though we may not hope to lay down a definition which shall clearly mark off the nation from all other human groups, we may usefully define the nation-state or nation in the most highly developed form that it has yet attained, and recognise that various peoples partake of the nature of, or approach the type of, the nation in so far as they exhibit something of its essential character.

It is perhaps hardly necessary to point out that it is only in the nation-state, or nation in the fullest sense of the word, that the state becomes identical with the nation, and that this identification has only been achieved in modern history by the growth among a few peoples of representative institutions and the democratic spirit.

In the work mentioned above Prof. Ramsay Muir writes—"What do we mean by a Nation? It is obviously not the same thing as a race, and not the same thing as a state. It may be provisionally defined as a body of people who feel themselves to be naturally linked together by certain affinities which are so strong and real for them that they can live happily together, are dissatisfied when disunited, and cannot tolerate subjection to peoples who do not share these ties."[1] The provisional definition has the merit of recognising that nationhood is essentially a mental condition and must be defined in psychological terms. The author goes on to inquire—"What are the ties of affinity which are necessary to constitute a nation?" He then considers the following conditions: (1) "occupation of a defined geographical area," (2) "unity of race," (3) "unity of language," (4) "unity of religion," (5) "common subjection, during a long stretch of time, to a firm and systematic government," (6) "community of economic

[1] *Op. cit.*, p. 38.

interest, with the similarity of occupations and outlook which it brings," (7) "the possession of a common tradition, a memory of sufferings endured and victories won in common, expressed in song and legend, in the dear names of great personalities that seem to embody in themselves the character and ideals of the nation, in the names also of sacred places wherein the national memory is enshrined." Of the last he says that it is "the most potent of all nation-moulding factors, the one indispensable factor"; thus showing his sense of the essentially psychological nature of nationhood. But of all the other six "factors" enumerated, he shows that they are unessential. After reaching this negative conclusion, that nationhood cannot be defined by any one of these marks or factors, he writes: "Nationality, then, is an elusive idea, difficult to define. It cannot be tested or analysed by formulæ, such as German professors love. Least of all must it be interpreted by the brutal and childish doctrine of racialism. Its essence is a sentiment, and in the last resort we can only say that a nation is a nation because its members passionately and unanimously believe it to be so. But they can only believe it to be so if there exist among them real and strong affinities; if they are not divided by any artificially maintained separation between the mixed races from which they are sprung; if they share a common basis of fundamental moral ideas, such as are most easily implanted by common religious beliefs; if they can glory in a common inheritance of tradition; and their nationality will be all the stronger if to these sources of unity they add a common language and literature and a common body of law. If these ties, or the majority of them, are lacking, the assertion of nationality cannot be made good. For, even if it be for the moment shared by the whole people, as soon as they begin to try to enjoy the freedom and unity which they claim in the name of nationality, they will fall asunder, and their freedom will be their ruin." In the last sentence the author clearly shows that the conclusion at which he seemed to have arrived, namely that "a nation is a nation because its members passionately and unanimously believe it to be so," is untenable. At the present time there are populations claiming the rights of nationality just upon this fallacious ground; a fact which illustrates the importance of achieving some satisfactory definition of nationhood. Indeed at the present moment, when Europe is being remoulded by the Paris Conference, the need for clear notions and some working definition of nationhood has acquired a most urgent importance. For, as our author remarks, "we say, loosely, that every nation has a *right* to freedom and unity," and the principle of "self-determination of nations" has become almost universally accepted as a kind of moral axiom of political justice; and this axiom is being applied to determine the political boundaries of the world now and for all time. Yet how can we hope to make a proper use of this principle, if we cannot define a nation, if a modern historian, who has devoted himself to the study of nationality, finds himself compelled in the year 1917 to give up the attempt to define the meaning of the term nation? For that is the issue of Prof. Ramsay Muir's interesting discussion. "We have not attained," he confesses, "in this discussion any very clear definition of nationality, or any very satisfactory test of the validity of the claims put forward for national freedom. We are not to base the doctrine of nationality upon abstract rights. We must recognise that there is no single infallible test of what constitutes a nation, unless it be the peoples' own conviction of their nationhood, and even this may be mistaken or based upon inadequate grounds."[1] And the dire consequences of this failure are made clear on the following page—"There seems no escape

[1] *Op. cit.*, p. 54.

from the conclusion that nationhood must mainly determine itself by conflict. That conclusion appears to be the moral of the history of the national idea in Europe." Which is as much as to say that, when any population declares itself to be a nation and claims the rights of nationhood, the Statesmen of the Paris Conference are to reply—"We do not know whether your claim is well-founded; for the historians and political philosophers cannot tell us the meaning of the word 'nation.' Go to and fight, and, if you survive, we shall recognise the *fait accompli* and hail you a Nation."

I have dwelt at some length on this perplexity of the historian, grappling with the task of defining nationhood because it illustrates so well a fact on which I wish to insist—namely, that it is not sufficient for the historian and the political philosopher to be willing to recognise the mental factors in the phenomena with which he deals. It is necessary to recognise that these factors are of overwhelming importance and that they cannot be satisfactorily dealt with by aid of the obscure and confused psychological concepts of popular thought and speech. We must recognise these political problems for what they truly are—namely, psychological through and through, and only to be attacked with some hope of success if we call to our aid all that psychological science can give us. This conclusion cannot fail to be unpalatable to very many workers in this field; for it implies that equipment for such work demands some additional years of preparatory study. But, it may fairly be asked, if the medical man must devote six years to the intensive study of the human body, before he is permitted to practise upon it, and even then without any scientific knowledge of the human mind, should not he who would practise upon the body politic, in which not merely the bodies but the minds of men interact in the most subtle and complex fashion, prepare himself for his exalted task by an even more extended course of study?

Prof. Ramsay Muir has the merit of recognising the essentially psychological nature of his problem; for his provisional definition (cited above) is wholly psychological, and he tells us that the essence of nationality is a sentiment; but he reveals the inadequacy of his psychological equipment by telling us in the same paragraph that its essence is a belief, the belief that they are a nation, passionately and unanimously held by the members of some group. If we look again at the list of seven proposed marks of nationhood, we shall see that they are rather of the nature of conditions favourable to the growth of nationhood; and, as we shall find, this list may be considerably enlarged. He comes nearest to the truth perhaps when he says "its essence is a sentiment." But he does not attempt to tell us what is the nature of this sentiment, nor even what is its object.

We may imagine a group of people of considerable magnitude, say the Mormons, or the Doukhobors, the Swedenborgians, or the Christian Scientists, withdrawing themselves to some defined territory, in order to form themselves into a nation; then, although each of the seven conditions enumerated by Prof. Ramsay Muir might be realised, and even though the community possessed the two conditions described by him as the essence of nationality—namely, a strong sentiment (presumably one of loyalty to the group) and a passionate belief in its nationality—it would, in the absence of other essential conditions, lamentably fall short of being a nation and would suffer the fate indicated; namely "as soon as they begin to try to enjoy the freedom and unity which they claim in the name of nationality, they will fall asunder, and their freedom will be their ruin."[1]

[1] *Op. cit.*, p. 51.

What, then, is the essential condition for lack of which any such people would fall short of nationhood? What is the factor which has escaped the analysis of Prof. Ramsay Muir? The answer must be—organisation; not material organisation, but *such mental organisation as will render the group capable of effective group life, of collective deliberation and collective volition.* The answer to the riddle of the definition of nationhood is to be found in the conception of the group mind. A nation, we must say, is a people or population enjoying some degree of political independence and possessed of a national mind and character, and therefore capable of national deliberation and national volition. In this and the succeeding chapters we have to examine the nature of such national mind and character, to give fuller meaning to these vague popular terms, and to study the way in which various conditions of national life contribute to their development.

Nationhood is, then, essentially a psychological conception. To investigate the nature of national mind and character and to examine the conditions that render possible the formation of the national mind and tend to consolidate national character, these are the crowning tasks of psychology.

Let me remind the reader at this point of the general sense of the words mind and character. The two words really cover the same content; when we speak of the individual mind or character, we mean the organised system of mental or pyschical forces which expresses itself in the behaviour and the consciousness of the individual man. Any such organised system has two aspects or sides which, though intimately related, may be considered abstractly as distinct—namely, the intellectual or cognitive aspect and the volitional, conative, or affective aspect. When we use the word "mind" in speaking of any such system, we give prominence to its intellectual side; when we say "character" we draw attention to its conative or affective side. The group mind of a nation is a mind in the sense that, like the mind of the individual, it is an organised system of mental or psychical forces; and, like the individual mind, it also has its intellectual and its affective sides or aspects. And this remains true whether or no there be any truth in that notion of the "collective consciousness" as a synthesis of minor consciousnesses which we have provisionally rejected;[1] that is to say, we accept unreservedly the notion of the collective mind, while suspending judgment upon the notion of "collective consciousness," until we shall find that this hypothesis is, or is not, required for the interpretation of the facts.

It will be observed that we are getting far away from the old-fashioned conception of psychology which limited its province to the introspective description of the contents of the individual's consciousness. The wider conception of the science gives it new tasks and new branches, of which the study of the national mind is one. Like the main trunk of psychology and most of its branches, this branch has to become an empirical science which shall take the place of what has long been regarded as a branch of speculative philosophy and pursued by the deductive *a priori* methods of philosophy. In this case the branch of philosophy in question has generally been called the Philosophy of History. It has been well said by Fouillée that the Philosophy of History of the past is related to the psychological social science, that is now beginning to take shape, as alchemy was related to chemistry, or astrology to astronomy. That is to say, it was a realm of obscure and fanciful ideas, of sweeping and ill-based assumptions and slipshod reasoning. It was an elaborate attempt "to

[1] Chapter II. On the question of the definition of the terms "mind" and "character" I would refer the reader to my *Psychology, The Study of Behaviour*, Home University Library.

lay the intellect to rest on a pillow of obscure ideas." The task of scientific analysis and research was avoided by bringing in, as the main explanatory principles or causal agencies, vaguely conceived entities regarded as presiding over the development of peoples—such entities as Providence, or the Destiny of nations, the *Genius* of a people, or the Instinct of a nation, the Unconscious Soul of a people, or the Spirit of the Age; and, when the problem was to account for some great secular change, for example, some change of national character, nothing was commoner than to appeal to *Time* itself, and thus to make of this most empty of all abstractions a directive agency and an all powerful cause of change. The strictly national gods of various nations were popular conceptions of this order; the gods who directly intervened in battles and enabled their chosen peoples to smite their enemies hip and thigh so that not one was left alive. Of this class the "good old German god" of the late German emperor was, it may be hoped, the last example.

In a less crude form similar hypotheses of direct supernatural intervention have been seriously maintained in modern times. Thus the poet Schiller argued as follows—"The individuals of whom a nation is composed are dominated by egoism, each seeking only his own good, yet their actions somehow secure the good of the whole; hence we must believe that the history of a people unrolls itself beneath the glance of a wisdom that looks on from afar, that knows how to control the ill-regulated caprices of liberty by the laws of a directing necessity and to make the particular ends pursued by individuals subservient to the unconscious realisation of a general plan."

In estimating the claims to consideration of a doctrine of this sort, we must put aside its deleterious moral effects, the fact that its acceptance would necessarily tend to weaken our sense of responsibility, to paralyse altruistic effort, and to justify purely egoistic conduct. We have to consider only its truth or probability in the light of history. When we do that, it appears merely as a fictitious solution of the larger problems of social science, a solution which may relieve us of the necessity of intellectual effort, but which brings no enlightenment and is supported by no serious argument. The one argument advanced is a libel on human nature; for it denies the reality and efficacy of the disinterested social efforts of the leaders of humanity, to which its progress has been in the main due; and it ignores the great mass of human activity due to the group spirit, with its fusion of egoistic and altruistic motives. Further, it ignores the fact that the history of the world is not merely the history of the rise of nations, but rather of the perpetual rise and *fall* of nations. When we are told that a power of this sort has constantly intervened in the course of history, and that the rise of peoples has been due to its guidance, we may fairly ask—Why has it repeatedly withdrawn its support, just when civilisation has achieved such a degree of development as might have rendered possible the flowering of all the finer capacities of human nature and the alleviation of the hard lot of the great mass of men? If the contemplation of the course of history compelled us to believe that such a power intervenes, we should certainly have to regard it as a malign power that delights in mocking human efforts by first encouraging and then bringing them to naught.

Very similar is the rôle in history assigned by von Hartmann to his "Unconscious." "It carries away the peoples that it dominates," says von Hartmann, "with a demoniac power towards unknown ends; it teaches them the way that they must take; though they often believe themselves to be marching towards a goal very different from that to which they are being conducted."

Others maintain that the great men of a nation, who are the principal agents in moulding its destiny, are in some mystical sense the products and expressions of the "unconscious soul" of the people, that they are the means by which its ideas are realised, through which they become effective; and they usually make the assertion, altogether unwarranted by history, that the moment of great need in the life of a people always produces a great man or hero to lead the people through the crisis. That is, or may appear to be, true of those peoples that have survived to pass into history. But what of those peoples that have gone down, leaving no trace of all their strivings, beyond some mounds of rubble, some few material monuments, or some strange marks on brick or stone or rock?

All such assumptions are the very negation of science. We have no right to appeal to such obscure and mystical powers, until by prolonged effort we shall have exhausted the possibilities of understanding and explanation in terms of known forces and conditions.[1]

On the other hand, a number of writers have sought to interpret the course of history and the rise and fall of nations in a more scientific manner; but most of these have studied some one aspect of national life, and have professed to find in that one aspect the key which shall unlock all doors and solve all problems. Thus some, adopting the notion of a variety of human races, each endowed with a certain peculiar and unalterable combination of qualities, seek to explain all history by the aid of biological laws, especially the Darwinian principles, as a struggle for survival between individuals and between races. Others, like Karl Marx and Guizot, see in economic conditions and the struggles between the social classes within each nation,

[1] Prof. Hans Driesch's conception of "super-individual entelechy" seems to be of this order, arrived at by the same line of reasoning. See *Science and Philosophy of the Organism*, Gifford Lectures, 1907.

the all-important factors. Others again, like Montesquieu and to some extent Buckle and more recently Matteuzzi, have seen in the influences of physical environment the key to the understanding of differences of national character and history; while others profess to have found it in differences of religious system, or of the forms of government and systems of laws. Others again, like Le Bon,[1] in a few dominant ideas which, they say, being possessed by any nation (or possessing a nation) determine its character and civilisation. All these are exaggerations of partial truths; and in opposition to all of them it must be laid down that the understanding of the *mind* of a nation is an indispensable foundation for the interpretation of its history.

Just as there are two kinds of psychology of individuals, so there are two kinds of psychology of peoples. There is the individual psychology which is primarily descriptive, which is the biography of persons, and whose aim is to impart an accurate conception of the general tendencies of a person and of the course of his development. And there is the psychology whose aim is to explain in general terms the conduct of individual men in general by the aid of conceptions and laws of general validity. The former, of course, was developed much earlier than the latter, which is in the main of quite modern growth. As this explanatory psychology develops, its principles begin to find application in the sphere of biographical or individual psychology, raising it also to the explanatory plane.

Just so there are two parallel kinds of psychology of peoples. There is the descriptive psychology of the tendencies of particular peoples, the biography of nations and peoples, which is what commonly is meant by "history"; and there is the psychology which seeks to explain in general terms how these tendencies arise, which seeks

[1] *Psychological Laws of the Evolution of Peoples.*

the general laws of which these diverse national tendencies are the outcome.

This last is the modern science which is beginning to take shape and to undertake the task so inadequately dealt with by the so-called Philosophy of History. It is essentially a branch, and by far the most important part, of Group Psychology.[1] Now individual psychology tends more and more to be a genetic psychology; because we do not feel that we really understand the individual mind, until we know how it has come to be what it is, until we know something of its development and racial evolution. Just so the explanatory psychology of peoples must be a genetic psychology. Here it differs from individual psychology in that the distinction between individual development and racial evolution disappears. For the national mind is a continuous growth; it is not embodied in a temporal succession of individuals, but in a single continuously evolving organism.

Nevertheless, we may with advantage consider separately (1) the nature of the general conditions necessary to the existence and operation of a national mind; (2) the processes of evolution by which such minds are formed and their peculiarities acquired. I propose to take up the former problem in the following chapter.

[1] As examples of the best work as yet accomplished in this immense and fascinating field, I would refer the reader to the books of M. Alfred Fouillée one of the most clear-sighted, judicious, and readable of modern philosophers, especially his *Psychologie des peuples européens*, his *Psychologie du peuple français*, and his *Science Sociale Contemporaine*.

CHAPTER VII

The Mind of a Nation

WE have prepared ourselves for the study of the national mind by our preliminary examination of the two extreme types of collective mental life, that of the quite unorganised group, the simple crowd, on the one hand, that of a very highly organised group, the army, on the other hand. We have seen that in the former type the collective actions imply a collective mental life much inferior, both intellectually and morally, to that of the average component individuals; and that in the other type they imply a collective mental life and capacities much superior to those of the average individual.

The mind of any nation occupies some intermediate position in the scale of which these are the extreme types; and it differs from both in being immensely more complex, and also in that the influence of the past dominates and determines to a much greater extent the mental life of the present.

The study we have already made of collective mental life will enable us to understand what we mean, or ought to mean, when we speak of national character. There are two senses in which this phrase is used, and they are often confused. On the one hand, the phrase may be used to denote the character of individuals who are taken to be typical representatives or average specimens of their nations. On the other hand, it may be taken to mean the character of the nation as a collective whole or mind

These two things are by no means the same; they are rather very different. We saw that this was true in the case of the crowd and also of the army; and it is true of the nation in a higher degree than of any other social aggregate, just because the influence of its past over its present is greater than in any of the others. It is in the second and preferable sense that Fouillée uses this expression. He writes—"The national character is not the simple sum of the individual characters. In the bosom of a strongly organised nation, there are necessarily produced reciprocal actions between the individuals which issue in a general manner of feeling, thinking, and willing very different from that of the individuals existing in isolation, or even from the sum or resultant of all the mental actions of isolated individuals. The national character is not simply the average type which one would obtain if one could imitate for minds the procedure adopted by Galton in the case of faces and so obtain a collective or generic image. The face which the process of compound photography produces exerts no action and is not a cause; while the national spirit does exert an effect which is different from all effects of individual minds; it is capable of exerting a sort of pressure and a constraint upon the individuals themselves; it is not only an effect, but is also in turn a cause; it is not only fashioned by individuals, it fashions them in turn. The average type of the Frenchman existing to-day, for example, does not adequately represent the French national character, because each people has a history, and ancient traditions, and is composed, as it is said, of the dead even more than of the living. The French national character resumes the physical and social actions that have been taking place through centuries, independently of the present generation, and imposes itself upon this generation through all the national ideas, the national sentiments and national institutions.

It is the weight of the entire history to which the individual is subjected in his relations with his fellow citizens. Just, then, as the nation, as a certain social group, has an existence different from (though not separable from) the existence of the individuals, so the national character implies that particular combination of mental forces of which the national life is the external manifestation."[1] That is a precise and admirable statement of what we are to understand by national mind and character.

We must now consider in turn the principal conditions of the existence of highly developed national mind and character, and first those which, as we have seen, are essential to all collective mental life.

A certain degree of mental homogeneity of the group, some similarity of mental constitution of the individuals composing it, is the prime condition. The homogeneity essential to a nation may be one of two kinds, native or acquired; both of these are usually combined, but one of them predominates in some nations, the other in others.

In considering racial or native homogeneity, we touch upon one aspect of a much disputed question, the influence of race on national character and history, in regard to which the greatest diversity of opinion has prevailed and still prevails. A correct estimate of this influence is of fundamental importance. I have stated elsewhere the view I take,[2] but we must consider the question more fully here. On the one hand are those who would explain all differences of national character and action, all success and failure of nations, as arising from racial composition. This view is the basis of much of the ill-founded national pessimism which, before the Great War, was widely prevalent among the peoples who speak the Romance or Latin languages and who are falsely called by these pessimists

[1] *Psychologie du peuple français*, p. 4, Paris, 1903.
[2] *Social Psychology*, p. 330.

the Latin races. It was also the foundation of that overweening national pride which has corrupted the German people and led them to disgrace and disaster; for, following Gobineau[1] and a host of his disciples, among whom H. S. Chamberlain is perhaps the most notorious, they had come to believe, against the most obvious and abundant evidence, that they were the purest representatives of a race from whose blood all great men and all good things have come, a race fitted by native superiority to rule all the peoples of the earth.[2]

On the other hand, popular humanitarianism would regard all men and all races as alike and equal in respect of native endowment; and we have seen so distinguished a sociologist as Durkheim denying any importance or influence to racial composition of a people. Many others put aside all explanations based on racial differences as cheap and meretricious means of avoiding difficulties. J. S. Mill, for example, wrote "Of all vulgar modes of escaping from the consideration of the effect of social and moral influences on the human mind, the most vulgar is that of attributing the diversities of conduct and character to inherent natural differences"; and Buckle, in his great work on the *History of Civilisation*, quoted this remark with cordial approval.[3]

Both these extreme views are false; the truth lies somewhere in the midst between them. At the time when Mill and Buckle wrote, biology and anthropology had not shown, as now they have, the enormous power of heredity in determining individual character and the great persistence of innate qualities through numberless generations.

[1] *Les inégalites des races humaines.*

[2] This fantastic doctrine has found its fullest expression in Chamberlain's work *The Foundations of the Nineteenth Century.*

[3] Other prominent exponents of this view are Mr. J. M. Robertson in his book *The Germans* and in his *Introduction to English Politics*, and M. J. Finot in his *Race Prejudice.*

Buckle especially overrated the power of physical environment, and Mill the power of education and of social environment, to change the innate qualities of a people; and it was this overestimation that led them, and leads others still, to underestimate the importance of racial composition. There are involved in this dispute two theses which are often confused together. When people speak of the influence of "race" on national character and institutions, they may, and sometimes do, mean by "race" the sum of innate inborn qualities or tendencies of the people at any given point of history. On the other hand, by influence of race they may mean the influence of the prehistoric races which have entered into the social composition of the nation—that is, those races from which its population is descended. Some authors mean to deny importance to race in both these senses; Buckle and Mill and Durkheim meant, I think, to deny it in both, because they believed that human nature is very plastic and easily moulded as regards its innate qualities by its environment; they believed that, if only a system of institutions, especially educational institutions, adapted to promote the intellectual and moral development of each generation of a people, can be established among it, then the influence of such institutions will so vastly predominate over that of innate qualities that these become a negligible quantity. From this it would follow that we should expect to see any two or more populations endowed with similar institutions form nations of similar character which will continue to develop along similar lines, except in so far as minor unessential differences of physical environment produce differences of modes of occupation, dress, food, and so forth.

This view of the insignificance of innate qualities was in harmony with, and was determined by, the dominant psychological doctrine of the time; the view which came down from Locke, according to which the mind of the new-born individual is a *tabula rasa*, entirely similar in all men, without specific tendencies and peculiarities of any importance, on which individual experience impresses itself, moulding all its development according to the principle of the association of ideas.

This doctrine, explicitly or implicitly adopted, has played a great part in determining British policy in its relations with British dependencies and their populations, notably India. It is a striking example of the way in which theory affects practice, and of the danger of our profound indifference to theory; we are influenced by it though we pretend to ignore it. It is well to make ourselves clear as to what theories we hold, even if we do not allow our practice to be governed by them exclusively.

There are commonly confused together, under the head of the influence of race on national character, three problems which must be disentangled.

(1) Are there differences of innate mental constitution between the various branches of mankind?

(2) If there are such differences, are these important for national life? Do they in any considerable degree determine national character? Or are they capable of being swamped and submerged and altogether overridden by the moulding influences brought to bear by environment on each generation?

(3) If such innate differences exist, what degree of permanence do they possess? Do they persist through thousands of years, in spite of vast changes of physical or cultural conditions? Or may they undergo considerable modification or complete transformation in the course of a few generations?

These are questions of fundamental importance. And they admit of no positive clean-cut answers at the present time. They offer vast fields for research, and only when prolonged research shall have been directed to them shall we be able to answer them positively.

In the past, since their importance could not be altogether overlooked, it has been usual to dispose of them by dogmatically asserting one extreme view and pouring scornful epithets upon the other extreme view. A principal task for science in its present stage is to define the questions clearly. It is not possible, perhaps, to keep them quite separate; for, if there are considerable differences of innate mental constitution, then their importance for national character must depend greatly upon their degree of permanence; and, again, there is the great difficulty of distinguishing between innate and acquired mental qualities in any individual and still more in groups.

Nevertheless, we may safely say that both extreme views in regard to race, the positive and the negative, are gross exaggerations, plausible only while we ignore one part of the evidence; the truth lies in between somewhere.

There can, I think, be no reasonable doubt that there are great differences between races, and that these may be, and in many cases have been, persistent through thousands of generations.

The recognition that the mind of the human infant is not a *tabula rasa*, but that its innate constitution comprises a number of instincts, specifically directed tendencies to thought, feeling, and action, prepares us to accept this view and gives us some basis for the definition of these differences. Whether all differences can be defined in such terms is a further problem. That they cannot be wholly defined in this way seems to be obvious, when we consider how quite specialised idiosyncrasies are transmitted in families through several generations, often with a leap across one generation, peculiarities of taste and feeling, of æsthetic endowment and temperament, abilities such as the musical, mathematical, and artistic.

When we compare widely different peoples such as the Negro, the White, and the Yellow, the fact of profound differences cannot be overlooked. These differences cannot be ascribed to the action of environment upon each generation. Perhaps the only differences of this kind which at present are accurately measurable are those of the size and form of the brain. The negro brain is decidedly smaller than that of the white and yellow races. And there are small but distinct differences of sensory endowment which are highly significant. For, if there are racial differences in these most fundamental and racially oldest endowments, we may expect still greater differences in the later evolved powers of the mind; although these are much more difficult to detect and define.

Still, the negro race wherever found does present certain specific mental peculiarities roughly definable, especially the happy-go-lucky disposition, the unrestrained emotional violence and responsiveness, whether its representatives are found in tropic Africa, in the jungles of Papua, or in the highly civilised conditions of American cities.

The Semitic stock again is one which, though widely scattered, seems to present certain constant peculiarities. And among closely allied branches of the white race of similar culture, we can hardly refuse to recognise innate differences. Differences of temperament are, perhaps, the clearest and the most generally recognised, even between peoples of allied stock and similar civilisation. Who can question that Irishmen in general are very different from Englishmen in temperament, that they are less phlegmatic, more easily moved to joy, or sorrow, or enthusiasm, more easily touched by poetry, have a more varied and lively emotional experience? That this is an innate racial difference seems clear; for it can be accounted for in no other way, and it obtains in some degree between all communities of similar racial stocks, in spite of similarities or differences of history and of present conditions. For example, similar differences, roughly definable as the difference between the so-called Celtic temperament and the Anglo-Saxon, seem evidently to obtain between the Breton and the Norman, who represent in the main the same two stocks.

And, even in intellectual quality, there appear to be not only differences of degree, but also differences of kind, inexplicable save as racial differences. The logical deductive tendencies of the French intellect and the empirical inductive tendency of the English, seem to be rooted in race; though here of course tradition accumulates and accentuates such differences from generation to generation.

But the best evidence of persistent innate differences is afforded by differences and similarities expressed in national life which cannot be accounted for in any other way. The innate differences and peculiarities of individuals are largely obscured by these national characteristics. And the more highly organised the collective life of any people, the more clearly will it express their racial qualities.

The social environment in a developed nation is in harmony with the individual innate tendencies, because in the main it is the natural outcome and expression of those tendencies. For, throughout the history of such a nation, the elements of its social environment—its customs, beliefs, institutions, language, its culture in general—have been slowly evolved under the steady pressure of the individual innate tendencies, which in each succeeding generation are the same. A part of this culture is of native origin; a part, in every European nation probably by far the larger part, is of foreign origin, and has been acquired by the acceptance of ideas and beliefs from without its borders, by the copying of institutions, customs, arts, from foreign models. In both cases the idea or custom or other cultural element only becomes embodied in the national culture through widespread or general imitation.[1]

In the case of elements of native origin, it is by imitation of the individuals of original powers of thought or feeling that the element becomes embodied in the national culture; in the case of foreign elements, by imitation of foreign models, acceptance of foreign ideas, through literature and personal contacts. In both cases, such general imitation will only take place when the culture-element in question is more or less congenial to the innate qualities of the bulk of individuals. All other novel elements will be ignored, or will fail to propagate themselves successfully; if they obtain a first footing, they will fail to pass beyond the stage of fashion into that of custom. And, when once accepted, the cultural element will usually undergo modification in the direction of more complete harmony with the innate tendencies; its less congenial features will be allowed to die out, its more congenial will be accentuated from generation to generation.

The social environment of any civilised people is, then, very largely the result of a long continued process of selection, comparable with the natural selection by which, according to the Darwinian theory, animal species are evolved; a constant favouring of certain elements, a constant rejection of others. We may in fact regard each distinctive type of civilisation as a species, evolved largely by selection; and the selective agency, which corresponds to and plays a part analogous to the part of the physical environment of an animal species, is the innate mental constitution of the people. The sum of innate qualities is the environment of the culture-species, and it effects a selection among all culture variations, determining the

[1] I here use this word in the large, loose, and convenient sense in which it is used by M. Tarde in his *Lois de l'imitation*. I have examined the nature of imitative processes more closely in my *Social Psychology*.

survival and further evolution of some, the extermination of others. And, just as animal species (especially men) modify their physical environment in course of time, and also devise means of sheltering themselves from its selective influence, so each national life, each species of civilisation, modifies very gradually the innate qualities of the people and builds up institutions which, the more firmly they are established and the more fully they are elaborated, override and prevent the more completely the direct influence of innate qualities on national life.

These principles are illustrated, perhaps, most clearly by the spread and modification of religious systems among peoples of different races. Take the case of the Moslem religion, which has gained acceptance among one-sixth of the population of the world in historic and in fact recent times, and is still spreading. The leading feature of this system is its acceptance of all that is and happens as being the will of God, the act of an entirely arbitrary, inscrutable, and absolutely powerful individual, before which men must simply bow without question or criticism; it is characterised by its simplicity and its fatalism. There seems good reason to believe that the tendency to unquestioning obedience to authority is a strong innate tendency of most Asiatics (except perhaps the Chinese and their relatives), far stronger than in most individuals of European peoples; for we see it expressed in many ways in their institutions and history, both of those who are and those who are not Moslems;[1] and Asiatic fatalism has, in fact, become proverbial. With the causes or origin of this innate quality we are not now concerned; but, accepting it as a fact, we may note that it is among Asiatics that Mohammedanism has secured the great mass of its converts; and that in India, in spite of many minor features

[1] Meredith Townsend regards this as one of the leading qualities of the peoples of India. See *Europe and Asia*, London, 1901.

that are opposed to the spirit of Hindooism, it continues to spread largely; while Christianity makes but little progress. Buddhism on the other hand has almost faded away, after an initial success in the country of its origin, but has continued to gain adherents and has become the dominant religion among the yellow peoples further east, in Burma, China, Thibet, Japan. The Moslem religion, having been thus accepted in virtue of the fact that its dominant tendency is in harmony with the strong innate tendency to unquestioning submission to a supreme will, then accentuates this tendency in all its converts, moulding their political relations to the same type, so that all recognise one earthly regent of God; and it has led to the almost complete suppression of any spark of the spirit of inquiry and scepticism that might otherwise display itself among these peoples.

Another good illustration of the fact is afforded by the distribution of the two great divisions of the Christian religion in Western Europe. Among all the disputes and uncertainties of the ethnographers about the races of Europe, one fact stands out clearly—namely, that we can distinguish a race of northerly distribution and origin, characterised physically by fair colour of hair and skin and eyes, by tall stature and dolichocephaly (i. e. long shape of head), and mentally by great independence of character, individual initiative, and tenacity of will. Many names have been used to denote this type, but the usefulness of most of them has been spoilt through their application to denote linguistic groups (e. g. Indo-Germanic, Aryan), and by the false assumption that linguistic groups are racial groups. Hence recently the term *Homo Europæus*, first applied by Linnæus to this type, has come into favour; and perhaps it is the best term to use, since this type seems to be exclusively European. It is also called the Nordic type.

The rest of the population of Europe, with the exception of some peoples in the extreme north and east of partly mongoloid or yellow racial origin, seems to be chiefly derived from two stocks. Of these, the one type, which occupies chiefly the central regions, is most commonly denoted by the name *Homo Alpinus;* the other, chiefly in the south, by the name *Homo Mediterraneus.* Both are of dark or brunette complexion and the principal physical difference between them is that the former, *H. Alpinus*, has a short, broad head (i.e. is brachycephalic) and also is of short stature; while the latter, *H. Mediterraneus*, is long-headed like the northern type and is perhaps taller than *H. Alpinus.* Mentally both these differ from the northern or European type in having less independence and initiative, a greater tendency to rely upon and seek guidance from authority.[1] Now we find that the distribution of the Protestant variety of Christianity coincides very nearly with the regions in which the fair type predominates; while in all other regions the Roman Catholic or Greek orthodox churches hold undisputed sway. North and South Germany illustrate the point. And Motley's account of the Netherlands shows how closely the line between Protestant Holland and Roman Catholic Belgium coincides with the line of racial division. We may note also that "Celts" of Ireland and Scotland early proved the superior strength of their religious tendencies by sending missionaries to England.

It would be absurd to hold that this coincidence is fortuitous. It is clearly due to the assimilation of the form of the religious and ecclesiastical system to the innate tendencies of the people. The northern peoples have given the system a turn compatible with the independence of spirit which is their leading racial quality; and among

[1] Cp. Ripley's *Races of Europe* and Prof. H. J. Fleure's *Human Geography in Western Europe*, London, 1919.

ourselves the tendency is apt to be pushed to an anarchical extreme in the rise of numerous small peculiar sects; this we must connect with the fact that the English represent in greatest purity the most independent branch of the Northern race.

The peoples among whom the other racial elements predominate have developed and maintained a religion of authority. And it is clear how, this differentiation having been achieved, either form of religion favours and accentuates in the peoples among whom it has become established the innate tendencies that have shaped it. The religion of authority tends, both by its general teachings and by the deliberate efforts of its official representatives, to suppress the spirit of independent thought and inquiry and action; the Protestant religion, relatively at least, favours the development of the independent tendencies of individuals. This is not to say that any individual is a Mohammedan or a Protestant, because he belongs to this or that race; that would be a parody of my statement. The form of each man's religious belief is, in the vast majority of cases, determined for him by the fact of his growing up within a community in which that form of belief prevails. My thesis is that in the main the racial qualities of each community have played a great part in determining which form of belief it shall accept. If the reader will reflect how, at the time of the Reformation, various communities hung for a time in the balance, he will see that the innate differences we have noted may well have played the determining rôle.

The same facts are illustrated by the political life of the European peoples. Only those among whom the northern race is predominant have developed individualistic forms of political and social organisation. Among the rest there appears clearly the tendency to rely upon the supreme authority of the state and to look to it for all initiative and guidance, a tendency to centralised and paternal administration; and this is the same, whether the external form of the political organisation be a monarchy or a republic. Thus France, in becoming a republic, did not overthrow the centralised system perfected by Henry IV, Louis XIV, Richelieu and Napoleon; for that system was congenial to the innate qualities of the mass of the people. It is clear that the centralised and therefore rigid system of government tends to accentuate, among the people subjected to it, their tendency to rely on authority and to repress individual initiative; while the other form, such as obtains in this country and still more in the United States of America, tends to the development of the initiative and independence of individuals, giving them free scope and throwing them upon their own resources. Among any people, an institution or other cultural element that has had a history of this kind will, then, cause a great development in the mass of individuals of just those innate or racial tendencies of which it is itself the slowly accumulated result or product.

If a nation is composed from stocks not too diverse, or if the original stocks have fused by intercrossing and have produced a fairly homogeneous people; and if this nation has enjoyed a long period of natural evolution undisturbed by violent influences from outside, conquests or invasions or immigrations on a great scale; then the social environment will have been brought in the main into harmony with the innate qualities of the people, and it will mould the individuals of each generation very strongly, accentuating and confirming those innate tendencies. This for two reasons. First, the social environment will be strongly organised and homogeneous; that is to say, the various elements, the beliefs, customs, institutions, and arts that go to compose it, will be in harmony with one another and of strongly marked character; and they will be almost universally accepted by that people as above criticism. Secondly, the institutions and customs have not to fight against the innate tendencies of the people in the formation of the adult minds, but co-operate harmoniously with them.

Now, when authors dispute over the question of the influence of race in determining the nation, they usually fail to distinguish clearly between the direct effects and the indirect effects of racial qualities.

Those who, like Mill, attribute to the social environment unlimited power of moulding individuals and who regard the influence of race as insignificant, are misled by the contemplation of such nations as we have been considering, the class of which our own is the most notable example, nations in which a strongly organised social environment makes in the direction of the innate tendencies. They overlook the fact that in any such nation the social environment, the body of institutions and traditions, is in the main the outcome and expression of these innate tendencies; they fail to see that the racial tendencies exert their strongest influence on national thought and action by means of the institutions, customs, and traditions on the growth of which they have exerted a constant directive pressure throughout many generations. In order to realise fully the influence of race, we must consider peoples whose culture and much else that enters into their social environment has been impressed upon them from without. We then see how little the social environment can accomplish in the moulding of a people, when it is not congenial to and in harmony with the racial tendencies.

The modern world contains certain instructive instances, of which Hayti is perhaps the most striking. There a circumscribed population of negro race has had a political and social and religious organisation and the elements of higher culture impressed upon it by Europeans, in the belief that it would be possible to construct a social environment which would mould the people. France, at a time of revolutionary enthusiasm for liberty, equality, and fraternity, withdrew from the island and granted the people self-government. The consequence has been a rapid relapse into barbarism and savagery of the worst kinds.[1]

It was the ignoring of the importance of race and the overestimation of the moulding influence of culture and institutions, eloquently voiced by Lord Macaulay, that led England eighty years ago to set out on the task of endowing the millions of India with British culture and institutions. The task has been pursued in a half-hearted manner only; but already we see some of the incongruity of the results of these efforts; and the best observers assure us that, were the task accomplished and the reins of a representative government left in native hands, it would be but a few years before the whole country would be reduced to a chaotic anarchic condition no better than that in which we found it. Others go further and assert with some plausibility that Western culture is positively injurious to the intellect and moral nature of Indians.[2]

In the Philippines the Americans seem to have applied similar mistaken ideas in a reckless fashion in the first years of their administration; with the result that, according to some accounts, they were in a fair way to plunge those islands into poverty and debt and chronic rebellion, while failing to secure affection, trust, or respect for themselves.

We must conclude, then, that innate mental constitution, and therefore race, is of fundamental importance in

[1] Cf. *The Black Republic*, by Sir Spencer St. John and *Where Black rules White*, by H. Hesketh Prichard.

[2] M. le Bon and more than one Indian civil servant in conversation.

determining national character, not so much directly as indirectly; for it gives a constant bias to the evolution of the social environment, and, through it, moulds the individuals of each generation. It will help to make clear the influence of innate qualities, if, by an effort of imagination, we suppose every English child to have been exchanged at birth for an infant of some other nation (say the French) during some fifty years. At the end of that period the English nation would be composed of individuals of purely French origin or blood; it would have the innate qualities of the present French nation; and the French nation would be, in the same sense, English. What would be the effect? Presumably things would go on much as before for a time. There would be no sudden transformation of our language, our laws, our religious or political institutions; and those who make little of the influence of race might point to this result as a convincing demonstration of the truth of their view. But gradually, we must suppose, certain changes would appear; in the course of perhaps a century there would be an appreciable assimilation of English institutions to those of France at the present day, for example, the Roman Catholic religion would gain in strength at the cost of the Protestant.

This view has been challenged and described as an extreme view.[1] But it is not. Both extreme opposite views continue to be maintained just because the importance of the indirect cumulative effect of innate qualities on culture is ignored. The innate qualities are of great importance, but only in the course of centuries can they exert their full effect on culture.

If then innate qualities have this importance, in what degree are they permanent? Here again two extreme views remain opposed to one another. Even as regards physical qualities this is still the case; and the problem is

[1] By G. Lowes Dickinson, *Hibbert Journal*, Jan., 1911.

much more difficult and at the same time infinitely more important as regards mental qualities. One reason for the belittling of innate qualities by Mill and Buckle, and for their overweening confidence in the power of institutions and environment, was the opinion generally prevailing in their time that, in so far as racial peculiarities exist, they can be modified and transformed in a few generations by physical and social environment.

But, when, under the influence of Weissman's theories, the majority of biologists came to the conclusion that acquired qualities are not transmitted, the position of the "race theorisers" was immensely strengthened. For selection, natural or social or artificial, remained as the only recognised cause of change of racial qualities; and, since it is clear that the development of civilisation tends to bring to an end the operation of natural selection, owing to the more efficient shielding of the weaker by the stronger members of societies, and since no other form of selection seems to have operated forcibly to change race qualities, it was inferred that race qualities endure throughout long ages with very little change.

Another revolution of opinion has had a similar effect. One of the old assumptions which seemed to justify the belief in rapid modifiability of race qualities was that the difference of culture between ourselves and our savage ancestors corresponds to, and is the expression of, an almost equally great difference of innate capacities, intellectual and moral. But this was in the main a misunderstanding. One well established fact suffices to show its improbability—namely, the larger size of the brains of Palæolithic men as compared with our own. Our superiority of civilisation is due to slow accumulation, each generation adding comparatively little to the mass of intellectual and moral tradition which it inherits and passes on to later coming generations. In so far as differences of cultural level are associated with differences of level of innate intellectual and moral qualities, cultural superiority must be regarded as the effect, rather than the cause, of innate mental superiority. There are strong grounds for holding that, in so far as Europeans are innately superior to Negroes, that superiority was achieved not by means of, and in the process of, the development of, civilisation; but rather before civilisation began; and that the principal mental differences of the various human stocks were, like their principal physical differences, produced in the course of the immensely long ages of human life that preceded the dawn of civilisation, or at any rate of history, ages compared with which the historic period is but a very brief span.

This view—namely, that there has been no great change, and certainly no great increase, of the mental powers of men during the historic period—was forcibly maintained by Dr. A. R. Wallace.[1] Wallace pointed to the pyramids of Egypt and other great achievements of earlier civilisations, such as writing, as evidence of the highly developed intellectual powers of men thousands of years before the Christian era. He concluded that the men of the early stone age were probably our equals, intellectually and morally, in respect to innate qualities.

If, then, so little change of man's mental constitution has been produced in the course of many thousand years, even though the growth of civilisation has so profoundly modified his mode of life and the nature of his pursuits, that is good evidence of the great persistency of racial mental qualities. But we have more direct evidence of their persistence. As Wallace points out, the negro and the yellow races are scattered over many parts of the earth, and, though these regions present great diversities of

[1] In the *Fortnightly Review*, Jan., 1910, and in his *Social Environment and Moral Progress*, London, 1913.

physical environment, men of either of the two races everywhere present the same mental peculiarities or strong similarities, for example, the Papuan and African and American Negroes. And the characteristic differences between the two races are not diminished even where, as in the islands of the far East, they have been subjected to the same physical environment and modes of life for long periods of time. In the eastern Archipelago, Papuans and Malayans occupying the same or adjoining islands are cited by Wallace as illustrating the persistence of racial mental differences; and I can bear out his remarks from my own observations in that region.

But we must beware of excess in the direction of the unalterability of race. The dogma of the non-transmissibility of acquired qualities is by no means established; it seems not improbable that mental acquisitions are so transmitted in some degree, though with only very slight effect in each generation. Even now, when the difficulties of the principle of transmission of acquired qualities are generally understood, almost all those who deal with the problem of the genesis of mental and physical peculiarities of races find themselves driven to postulate the principle in order to explain the facts. And this in itself constitutes evidence of a certain value in support of the validity of the principle.

Again, we must beware of assuming that there are no selective processes operating among us. Although natural selection may be almost inoperative, there may well be at work other forms of selection, social selections; and these are specially powerful amongst populations of blended stocks.

Summing up on the durability of racial peculiarities, we may say that racial qualities are extremely persistent; but that, nevertheless, they are subject to slow modifications when the conditions of life are greatly changed, as by emigration, or by changes of climate, or by social revolutions, and especially among populations of mixed origin.

To return now to the question of mental homogeneity of a population as a condition of national character and collective mental life. Purity of race is the most obvious condition of such homogeneity; but few, if any, nations that have attained any high level of civilisation have been racially homogeneous; probably for the simple reason that the civilisation of such a nation would crystallise at an early stage into rigid forms which would render further progress impossible. This has been the fate of most civilisations of the past; as Walter Bagehot put it, their cake of custom has so hardened as to become brittle, incapable of partial modification and growth, so that, like a crystal, it must either resist completely every modifying influence or be shattered irretrievably.[1]

Certainly none of the European nations are racially homogeneous. Nevertheless, some of them approach homogeneity of innate qualities, or, rather, the degree of heterogeneity is much less in some than in others. Consider the case of England. Before the Anglo-Saxon invasion the population consisted in all probability of a mixture of the northern fair race with a darker race, probably that of *H. Mediterraneus*, in some proportion that we cannot determine, with small islands of *H. Alpinus* or of stocks formed by an earlier blending of this with the Nordic race. The Anglo-Saxon invasion brought great numbers of the pure representatives of the Northern race of closely allied stocks; and these did not confine themselves to any one region, but, entering at many points of the south and east coasts, diffused themselves throughout almost all England, imposing themselves as masters upon those Britons whom they did not drive out. Ever since that time a crossing of the stocks has been going on freely,

[1] *Physics and Politics.*

little hindered by differences of area, language, law, or custom. And, with the exception of small numbers of the northern stock, Danes and Normans, the population has not received any considerable additions since the Saxon invasion.

Now it has been shown by a simple calculation that, given three generations to the century, each one of us might claim ten million ancestors in the year 1000 A.D.; while in the fifth century, when this process of intermarriage began, the number would be enormous, some thousands of millions; that is, if consanguine marriages had never taken place. These figures make it clear that, in any mixed population in which intermarriage takes place freely, the two or more stocks must, after a comparatively brief period of time, become thoroughly blended, on one condition—namely, that the cross between the pure stocks is a stable stock, fertile *inter se* and with both the parent stocks. There seems to be no doubt that this was the case with the British and the Anglo-Saxon stocks, and that the English form now a stable new subrace, or secondary race, in which the qualities of the northern race predominate. The subrace may be regarded as innately homogeneous in fairly high degree; not so homogeneous as a people of unmixed racial origin, or one formed by a blending of more remote date, but more so than most of the European nations. This is the sense in which we must understand the word race, in discussing the influence of race upon national character.

In most of the European countries the original mixture of races has been greater and the degree of blending less intimate. Thus France has the three stocks, *H. Europæus*, *Alpinus*, *Mediterraneus*, all largely represented; but they have remained in some degree geographically separated in three belts running east and west.[1] Hence

[1] Cp. Ripley, *op. cit.* and Fleure, *op. cit.*

there are greater innate mental differences between Frenchmen than between Englishmen. Nevertheless the strength of the Roman civilisation of Gaul sufficed to abolish differences of language and institutions and to assimilate the later coming Northmen, Franks, and Normans; while the centralised system of administration, established in accordance with the innate tendencies of the major part of the population, has completed the work of a long series of national wars, and has produced a firmly united nation, bound by common traditions and moulded by common institutions. The greater centralisation of France seems to have compensated for the less degree of innate uniformity, so that the French people is hardly, if at all, less truly a nation than our own.

In our own nation one racial cleft still remains. The Irish have never undergone that intimate mixture and blending with the Anglo-Saxon stock which has produced the English subrace; and so they remain an element which seriously disturbs the harmony of the national mind. And the same is perhaps true in a less degree of the Welsh people. On the other hand, the Scottish people, although they enjoyed their independent system of government for much longer periods than the Irish and Welsh and have a system of laws and customs differing in many respects from the English, and indeed may be said to have achieved a considerable degree of independent nationhood, have nevertheless become thoroughly incorporated in the British nation; for in the main mass of the Scotch the same northern race is the greatly predominant element.

But it is not till we consider such a country as Austria-Hungary that we see the full importance of homogeneity of a people for the development of a national mind. There several races and subraces, one at least with a strong yellow strain, are grouped together under one flag; but they remain separated by language and by distribution and by tradition, and, therefore, are but little mixed and still less blended. Under such conditions a national mind cannot be formed. The elements of different racial stock threaten to fall apart at any moment.[1]

Going further afield, contrast India with China, two regions geographically comparable in area and in density of population and in other ways. The population of China is the most racially homogeneous of all large populations in the world. Hence an extreme uniformity of culture and social environment, which still further accentuates the uniformity of mental type. Hence, in spite of the imperfection of means of communication, we find great political stability and a considerable degree of national feeling, likely to be followed before long by harmonious national thought and action on the part of this vast nation. The one great distracting and disturbing factor in the life of China has been the intrusion of the Manchus, a people of somewhat different race and traditions.

On the other hand, India is peopled by many different stocks, and, although these are geographically much mixed, they are but very little blended, owing to the prevalence from early times of the caste system. The light coloured intellectual Brahman lives side by side with small black folk, as different physically and mentally as the Englishman and the Hottentot; and there are also large numbers of other widely differing racial stocks, including some of yellow race. Hence an extreme diversity of social environment, save in the case of the Moslem converts, who, however, being scattered among the rest, do but increase the endless variety of custom, creed, and social environment. Hence the people of India have never been bound together in the slightest degree, save purely externally by the power of foreign conquerors, the Moguls

[1] This was written in 1910, and now in 1919 the dissolution which was so obviously impending is an accomplished fact.

and the British; and hence, even though nations have begun at various times to take form in various areas, as e. g. the Sikh nation, they have never achieved any high degree of permanence and stability and are restricted in area and numbers.

Now, let us consider for a moment an apparent exception from the conclusion to which the foregoing argument seems to point—namely, that homogeneity of innate qualities is the prime condition of a developed and harmonious national life.

The most striking exception is afforded by the people of the United States of America, or the American nation. There we see a great area populated by immigrants from every part and race of Europe in times so recent that, although they are pretty well mixed, they are but little blended by crossing; a considerable part of the population still consisting of actual immigrants and their children. Here, then, there can be no question of any homogeneity as regards innate mental qualities. Nevertheless, the people is truly a nation and, perhaps, further advanced in the evolution of national consciousness, thought, and action than many other of the civilised peoples. This we must attribute to homogeneity of mental qualities which is in the main not innate but acquired, a uniformity of acquired qualities, especially of all those that are most important for national life.

Following Münsterberg's recent account of *The Psychology of the American People* we may recognise as individual characteristics, almost universally diffused, a spirit of self-direction and self-confidence, of independence and initiative of a degree unknown elsewhere, a marvellous optimism or hopefulness both in private and public affairs, a great seriousness tinged with religion, a humorousness, an interest in the welfare of society, a high degree of self-respect, and a pride and confidence in the present and still more in the future of the nation; an intense activity and a great desire for self-improvement; a truly democratic spirit which regards all men (or rather all *white* men) as essentially or potentially equal, and a complete intolerance of caste.

Such high degree of acquired homogeneity of individual qualities seems to be due in about equal parts to uniformity of social and of physical environment, both of which make strongly in the same direction. The physical environment consists in a great and rich territory, still only partially developed, a fairly uniform climate, and a uniformity of the physical products of human labour resulting from the immense development of the means of communication. The importance of the physical uniformity we may realise on reflecting that the one great divergence of physical conditions, the sub-tropical climate of the southern States, gave rise to the one great and dangerous division of the people which for a time threatened the harmonious development of the national life; that is to say, the civil war was due to the divergence of the social system and economic interests of the southern States resulting from their sub-tropical climate.

The uniformity of social environment we must ascribe, firstly and chiefly, to the fortunate circumstance that the first immigrants were men of one well marked and highly superior type, men who possessed in the fullest measure the independence of character and the initiative of the fair northern race, and who firmly established the superior social environment of individualistic type that had been gradually evolved in England. Secondly, to the fact that the peopling of the whole country has taken place by diffusion from this strongly organised initial society; its institutions and ideas, especially its language, its political freedom, its social seriousness, being carried everywhere. Thirdly, to the fact that the country was just such as to give the greatest scope to, and so to develop, these innate tendencies of the earliest settlers and their successors. Fourthly, to the fact that the great diffusion of the population of mixed origin has only taken place since the means of communication have become very highly developed. Consider, as one example of the effects of the ease of communication between all parts, the influence of the American Sunday newspapers. These papers are read on an enormous scale all over the continent; and the bulk of the contents of those published in different places is identical, being prepared and printed in New York, or other great city, and then sent out to be blended with a little local matter in each centre of publication; thus every Sunday morning vast numbers are reading the same stuff. Lastly, it must be added, it is largely due to the fact that in the main the population has been recruited by those elements of different European peoples who shared in some degree the leading tendencies of the American character, independence, initiative, energy, and hopefulness; for it is only such people who will tear themselves from their places in an old civilisation and face the unknown possibilities of a distant continent. In spite of an increasing proportion of emigrants of a rather unlike type from southeastern Europe, there seems good ground for hope that these factors will continue to secure a sufficient uniformity of acquired qualities, until the diverse elements shall have been fused by intermarriage to a new and stable subrace, innately homogeneous.

The Americans are, then, no exception to the rule that the evolution of a national mind presupposes a certain considerable degree of homogeneity of mental qualities among the individuals of which the nation is composed. They merely show that, under peculiarly favourable physical and social conditions, a sufficient degree of such homogeneity may perhaps be secured in spite of considerable racial heterogeneity. But the favourable issue of the vast experiment is not yet completely assured.

There remains in the American people one great section of the population, namely the Negroes and the men of partly Negro descent, whose innate qualities, mental and physical, are so different from those of the rest of the population, that it seems to be incapable of absorption into the nation. This section remains within the nation as a foreign body which it can neither absorb nor extrude and which is a perpetual disturber and menace to the national life. The only hope of solving this difficult problem seems to lie in the possibility of territorial segregation of the coloured population in an area in which it might, with assistance from the American people, form an independent nation. At present it illustrates in the most forcible manner the thesis of this chapter.

The geographical peculiarities of the country inhabited by a nation may greatly favour, or may make against, homogeneity, in so far as this depends on acquired interests and sentiments.

The division of the territory occupied by a nation by any physical barrier makes against homogeneity and therefore against national unity; whereas absence of internal barriers and the presence of well-marked natural boundaries afford conditions the most favourable to homogeneity.

Almost all the great and stable nations have occupied well-defined natural territories. In Great Britain and Japan the national spirit is perhaps more developed than elsewhere. How much does Great Britain or Japan owe this to the insular character of its territory, which from early days has sharply marked off the people from all others, making of them a well-defined and closed group, within which free intermarriage has given homogeneity of innate qualities, and within which a national culture has grown up undisturbed; so that by mental and physical type, and by language, religion, tradition, and sentiment, the people are sharply marked off from all others, and assimilated to one another!

A unitary well-defined territory of well-marked and fairly uniform character tends to national unity, not only through making the community a relatively closed one, but also by aiding the imagination to grasp the idea of the nation and offering a common object to the affections and sentiments of the people.

Contrast in this respect the physical characters of England and Germany. The boundaries of the latter are almost everywhere artificial and arbitrary and have fluctuated greatly. It would be impossible for a poet to write of Germany as Shakespeare wrote of England:——

> This fortress built by Nature for herself
> Against infection and the hand of war;
> This happy breed of men, this little world,
> This precious stone set in the silver sea,

and all the rest of that splendid passage. France, Spain, Italy, Greece, Denmark, Scandinavia, are all more fortunate than Germany or Austria in this respect; and the lack of such natural boundaries has been in the past, and threatens to be in the future, a source of weakness to the German nation. We may, I think, not improbably attribute, in part at least, to this circumstance a peculiarity often noticed in German emigrants—namely, that they rapidly become denationalised and assimilated by the peoples among whom they settle, and that an Americanised German, for example, has often less sympathetic feeling for Germany than a foreigner. For, owing largely to the lack of natural boundaries and the consequent fluctuations that have occurred, and the mingling and blending with other peoples, Germany is a less clear-cut conception than Great Britain or France; to be a German is something much less definite than to be an Englishman or a Japanese, or even a Frenchman or a Spaniard.[1] And the presence or lack of definite natural territorial boundaries operates in a similar way through many centuries, determining on the one hand historical continuity to a people as a whole, or on the other hand breaches of historical continuity.

The United States of America afford a fine example of the binding influence of a well-defined territory; for here the effect is clearly isolated from racial factors and from slowly accumulated tradition. The Monroe doctrine is the outward official expression of this effect. The private individual effect is a sense of part ownership of a splendid territory with a great future before it. And we are told, I believe truly, that this sense is very strong and very generally diffused even among immigrants; that it inspires an unselfish enthusiasm for the work of developing the immense resources of the country; that this is the idealistic motive of much of the intense activity which we are apt to ascribe to the love of the "almighty dollar"; and that it is one of the main causes of the rapid assimilation of immigrants to the national type of mind.

The Chinese nation, again, owes its existence and its homogeneity of mental and physical type to geographical unity. Roughly, China consists of the basins of two immense rivers, not separated from one another by any great physical barrier, but forming a compact territory well marked off save in the north. It comprises no such partially separate areas as in Europe are constituted by Spain, or Italy, or Greece, or Scandinavia, or even France;

[1] The great myth of the racial unity and superiority of the German people which we have noticed above, has been cultivated and propagated, with elaborate disregard for fact by the German State and its henchmen in the universities and elsewhere, in a deliberate effort to remedy by art the lack of natural boundaries and of true national homogeneity.

almost all parts are well adapted for agriculture. Hence, largely, the national unity and the national sentiment which have long existed, and possibly a latent capacity for national thought and action.

Perhaps the most striking instance of all is ancient Egypt. There, in the long strip of land rendered fertile by the waters of the Nile, a people of mixed origin was long shut up and isolated; there all men felt their immediate dependence on the same great powers, the great river which once a year overflows its banks, and the scorching sun which passes every day across a cloudless sky. There all men looked out on the same unvarying and unvaried landscape, hoped and feared for the same causes, suffered the same pains, prayed for the same goods. There was formed one of the most stable and enduring of nations, whose uniform culture certainly bears the impress of the uniform monotonous physical environment.[1]

The other way in which physical environment affects homogeneity is by determining similarity or difference of occupations and, through them, similarities or differences of practical interests and of acquired qualities. So long as such differences are determined in many small areas, the result is merely a greater differentiation of the parts, without danger to the unity of the whole nation. But, when the physical differences divide a whole people into two or more locally separate groups differing in occupation and interests and habits, they endanger the unity of the whole. There are to-day many countries in which the distribution of mineral wealth is exerting an influence of this sort, giving rise to the differentiation of an industrial area from agricultural areas and a consequent divergence of interests and of mental habits; notably South Africa, Spain, and Italy.

[1] In his *Works of Man* Mr. March Philips shows clearly the influence of the Egyptian landscape upon the arts of sculpture and architecture.

Great Britain is fortunate in this respect also. Its geological formation presents on a small scale all the principal strata from the oldest to the most recent, a fact which secures great diversity within a compact area, an area too compact to allow of divergences of population being produced by differences of geological formation; so that it enjoys the advantages of diversity without its drawbacks. Although a certain degree of differentiation between north and south may be noted, it is not sharp or great enough to be dangerous. But let us imagine that coal and iron had been confined to Scotland. Would there be now the same harmony between the two countries as actually obtains? The United States of America affords a good illustration of this principle, as I have already pointed out; the subtropical climate of the southern States gave rise to a differentiation of occupations, and consequently of ideas and interests and sentiments, which was almost fatal to the unity of the nation. A similar differentiation between the agricultural west and the industrial and commercial east seems to be the greatest danger to the future unity of the nation; and the same may be said of the Canadian people.

Ireland illustrates well the effects of both kinds of physical influence. The Irish Channel has perpetuated that difference of race and consequent difference of religion which, but for it, would probably have been wiped out by free intermarriage; while the lack of coal and iron in the greater part of the country has prevented the spread of industrialism, and has thus accentuated the difference between the Irish people and the English. And it is obvious that among the Protestants of Ulster the accessibility of coal and iron, maintaining a divergence of occupations and of interests which prevents racial and cultural blending, perpetuates the racial and traditional differences between them and the rest of the population.

CHAPTER VIII

Freedom of Communication as a Condition of National Life

LET us consider now very briefly in relation to the life of a nation a second essential condition of all collective mental life—namely, that the individuals shall be in free communication with one another. This is obviously necessary to the formation of national mind and character. It is only through an immense development of the means of communication, especially the printing press, the railway and the telegraph, that the modern Nation-State has become possible, and has become the dominant type of political organism. So familiar are we with this type, that we are apt to identify the Nation and the State and to regard the large Nation-State as the normal type of State and of Nation, forgetting that its evolution was not possible before the modern period.

In the ancient world the City-State was the dominant type of political organism; and to Plato and Aristotle any other type seemed undesirable, if not impossible. For they recognised that collective deliberation and volition are essential to the true State. Aristotle, trying to imagine a vast city, remarks—"But a city, having such vast circuit, would contain a nation rather than a state, like Babylon." The translator there uses the word "nation," not in the modern sense, but rather as we use "people" to denote a population of common stock not organised to form a nation. The limits of the political organism capable of a collective mental life were rightly held to be set by the number of citizens who could live so close together as to meet in one place to discuss all public affairs by word of mouth.

The great empires of antiquity were not nations; they had no collective mental life. Although the Roman Empire, in the course of its long and marvellous history, did succeed in generating in almost all its subject peoples a certain sentiment of pride in and attachment to the Empire, it cannot be said to have welded them into one nation; for, in spite of the splendid system of roads and of posting, communication between the parts was too difficult and slow to permit the reciprocal influences essential to collective life. As in all the ancient empires, the parts were held together only by a centralised, despotic, executive organisation; there was no possibility of collective deliberation and volition.[1]

All through history there has obviously been some correlation between the size of political organisms and the degree of development of means of communication. At the present time those means have become so highly developed that the widest spaces of land and sea no longer present any insuperable limits to the size of nations; and the natural tendency for the growth of the larger states at the expense of the smaller, by the absorption of the latter, seems to be increasingly strong. It seems not unlikely that almost the whole population of the world will shortly be included in five immense States—the Russian or Slav, the Central European, the British, the American, and the Yellow or East Asiatic State. The freedom of communication between the countries of Europe is now certainly sufficient to allow of their forming a single nation,

[1] Cf. Sir S. Dill's *The Roman Empire from Nero to Augustus*, London, 1905.

if other conditions, such as diversities of racial type and of historical sentiments, would permit it.

Although, then, the platform and the orator and the assembly remain important influences in modern times, it is primarily the telegraph, wireless telegraphy, the printing press, and the steam engine, that have rendered possible the large modern nations; for these have facilitated the dissemination of news and the expression of feeling and opinion on a large scale, and the free circulation of persons.[1]

Without this freedom of communication the various parts of the nation cannot become adequately conscious of one another; and the idea of the whole must remain very rudimentary in the minds of individuals; each part of the whole remains ignorant of many other parts, and there can be no vivid consciousness of a common welfare and a common purpose. But, more important still, there can be none of that massive influence of the whole upon each of the units which is of the essence of collective mental life. Of these means of reciprocal influence the press is the most important; though, of course, its great influence is only rendered possible by the railway and the telegraph.

Hence we find that it is as regards the press that Great Britain and America differ most markedly from such states as Germany, and still more Russia. To an Englishman or American the meagre news-sheets which in Germany take the place of our daily and weekly press bring a shock of astonishment when he first discovers them; and that astonishment is not diminished when he finds that the best people hardly trouble to look at them occasionally.[2]

[1] Since these lines were written a new mode of rapid locomotion, namely the aerial, which has resulted from the invention and rapid development of the internal combustion engine, threatens to eclipse all others in its effects upon the organisation of the world.

[2] This state of affairs has no doubt been considerably altered during the great war; the political education of Germany, a painful but salutary process, is progressing rapidly.

It is interesting to note how the general election of January, 1910, illustrated the importance of improved means of communication. It was found that the number of citizens voting at the polls was a far larger proportion of those on the register than at any previous election; and, in this respect, the election was a more complete expression of the will of the people than any preceding one. This seems to have been due to the use of the motor-car, at that time the latest great addition to our means of communication.

The modern improvements of means of communication tend strongly to diminish the importance of the geographical factors we considered in the foregoing chapter; for they practically abolish what in earlier ages were physical barriers to intercourse; they render capital and labour more mobile; and they make many forms of industry less dependent upon local physical conditions and, therefore, less strictly confined by geographical factors. As instances of important developments of this order in the recent past or near future, the reader may be reminded of the railway over the Andes between Chile and Argentina, the tunnels through the Alps, the Channel tunnel, the Siberian railway, the Suez and Panama canals, the Cape to Cairo railway, and, above all, aerial transport. All these make for free intercourse between peoples.

Easy means of communication promote development in the direction of the organic unity of a nation in another way—namely, they promote specialisation of the functions of different regions; they thus render local groups incapable of living as relatively independent closed communities; for they make each local group more dependent upon others, each upon all and the whole upon each; hence they develop the common interest of each part in the good of the whole.

This influence already extends beyond national groups and it had been hoped that its further growth was about to render war between nations impossible.[1] To-day England is contemplating a task never before attempted, the fusing into one nation of the peoples of the mother-country and her distant colonies. Whether or no she will succeed depends upon whether the enormously increased facilities of communication can overcome the principal effects of physical barriers that we have noted—namely, lack of intermarriage and divergence of occupations, with the consequent divergence of mental type and interests. The task is infinitely more difficult than the establishment of such an Empire as the Roman; not because the distances are greater, but because the union must take the form of nationhood, because it must take the form of a collective mind and not that of a merely executive organisation. But, in the considerations which have shown us that membership in and devotion to a smaller group is by no means adverse to membership in and devotion to a larger group, we have ground for believing that the task is not impossible of achievement.

The slow rate of progress towards nationhood of such peoples as the Russian and the Chinese has been largely due to lack of means of free communication between the parts of these countries. On the introduction of improved communications, we may expect to see rapid progress of this kind; for many of the other essential conditions are already present in both countries.

The fact that in the Nation-State the communications between individuals and between the parts of the whole are in the main indirect, mediated by the press, the telegraph, and the printed word in general, rather than by voice and gesture and the other direct bodily expressions of thought and emotion, modifies the primary manifestations of group life in important ways which we must notice in a later chapter.

[1] Cf. N. Angell, *The Great Illusion.*

CHAPTER IX

The Part of Leaders in National Life

WE turn now to a third very important condition of the growth of the national mind, one which also has its analogue in both the crowd and the army. A crowd always tends to follow some leader in thought, feeling, and action; and its actions are effective in proportion as it does so. To follow and obey a leader is the simplest, most rudimentary fashion in which the crowd's action may become more effective, consistent, intelligent, controlled. Not any one can be such a leader; exceptional qualities are necessary. In every army the importance of leadership is fully recognised. A hierarchy of leaders is the essence of its organisation. In the deliberately organised army, the appointment of leaders is the principal and almost the sole direct means taken by the State to organise the army. Everything is done to give to the leaders of each grade the greatest possible prestige, especially by multiplying and accentuating the distinctions between the grades. Though much can be accomplished in this way, unless the men chosen as leaders have in some degree the superior qualities required by their position in the hierarchy, the whole organisation will be of little value.

The same is true in much higher degree of nations. If a people is to become a nation, it must be capable of producing personalities of exceptional powers, who will play the part of leaders; and the special endowments of the national leader require to be more pronounced and exceptional, of a higher order, than those required for the exercise of leadership over a fortuitous crowd.

Such personalities, more effectively perhaps than any other factors, engender national unity and bring it to a high pitch. There are regions in which the other main conditions of national unity have long obtained, but which have failed to become the seat of any enduring nation. Although the greater part of Africa, perhaps the richest continent of the globe, has been in the possession of the Negro races during all the ages in which the European, Asiatic, and American civilisations were being developed, those races have never founded a nation. Nevertheless many, perhaps most, Negroes are capable of acquiring European culture and of turning it to good account. And, when brought under the influence of Arabs or men of other races, they have formed rudimentary nations.[1] The incapacity to form a nation must be connected with the fact that the race has never produced any individuals of really high mental and moral endowments, even when brought under foreign influences; and it would seem that it is incapable of producing such individuals; the few distinguished Negroes, so called, of America—such as Douglass, Booker Washington, Du Bois—have been, I believe, in all cases mulattoes or had some proportion of white blood. We may fairly ascribe the incapacity of the Negro race to form a nation to the lack of men endowed with the qualities of great leaders, even more than to the lower level of average capacity. On the other hand, there is at least one people which, in the absence of every other condition, has continued to retain something of the character of a nation for many generations—namely, the Jews. The Jews are not even racially homogeneous, and they are scattered through all the world under the most varied

[1] Incipient nations have appeared where the Bantu stock has produced occasionally great warrior chiefs such as Chaka and Cetewayo.

physical conditions; yet the influence of a succession of men of exceptional power, Moses and his successors the prophets, all devoted to the same end—namely, the establishment of the Jewish nation and religion—has lived on through many generations and still holds this people together, marking it off from all others.

This is an extreme instance. But another almost equally striking case is that of the Arab nation, which has owed its existence to one man. The Arab nation was made by the genius of Mahomet, who welded together, by the force of his personality and the originality and intensity of his religious conviction, the warring idolatrous clans of Arabia. Until his advent these had been a scattered multitude, in spite of racial and geographical uniformity, geographical isolation, and fairly free intercommunication. We have here one of the purest, clearest instances of the effect of great personalities in furthering nationhood; for there seems to be no reason to believe that, if Mahomet had not lived, any such development of the Arabian people would have taken place.

M. le Bon has produced a curious piece of evidence bearing on this question. He has measured the cranial capacity of a great number of skulls of different races, and has shown that any large collection of skulls from one of the peoples who have formed a progressive nation invariably contains a certain small number of skulls of markedly superior capacity, implying exceptionally large brains; while any similar collection of skulls from one of the unprogressive peoples, like the Negro, differs, not so much in the smaller average size of the brain, as in the greater uniformity of size, that is to say, the absence of individuals of exceptionally large brains.[1] He, rightly, I think, sees in the absence of such individuals a main condition of the unprogressive character of these races; and, in the ex-

[1] *Psychological Laws of the Evolution of Peoples.*

ceptionally large brains produced among the other peoples a main condition of their progress.

These indications are borne out by a review of the history of any nation that has achieved a considerable development. Every such people has its national heroes whom it rightly glorifies or worships; for to them it owes in chief part its existence.

To them also it owes in large measure the forms of its institutions, its religion, its dominant ideas and ideals, its morals, its art and literature, all that of which it is most proud, all its victories of peace as well as of war, the memory of which and the common pride in which is the strongest of all national bonds.[1]

Who can estimate the enormous influence of Confucius and Lao-Tse in moulding and rendering uniform the culture of China? The influence of single individuals has undoubtedly been greater in the early than in the later stages of civilisation; for there was then a more open field, a virgin soil, as it were, for the reception of their influence. In the developed nation the mass of accumulated knowledge and tradition is so much greater, that the modifications and additions made by any one man necessarily are relatively small.

The leading modern nations owe their position to their having produced great men in considerable numbers; for that reason also no one man stands out so prominently as Mahomet or Confucius or Moses. Nevertheless their existence can in many cases be traced to some few great men. Would Germany now be a nation, but for Frederick the Great and Bismarck? Would America, but for

[1] I do not propose to examine in this book the much discussed question—Are the leaders of a nation to be regarded as produced by the nation according to the general laws of biology and psychology, or as given to them by some supernatural process? This question belongs to a branch of Social Psychology which is not included in the volume.

Washington, Hamilton, and Lincoln? Would Italy, but for Garibaldi and Mazzini and Cavour? How greatly is the unity of national spirit and tradition among Englishmen due to the great writers who have produced the national literature, and to the great statesmen and soldiers and sailors who have given her a proud position in the world! What would England be now if Shakespeare, Newton and Darwin, Cromwell and Chatham, Marlborough and Nelson and Wellington had never been born?

And it is not only the men of great genius who are essential to the modern nation, but also men of more than average powers, though not of the very highest.

Let us try to imagine the fifty leading minds in each great department of activity suddenly removed from among us. That will help us to realise the extent to which the mental life of the nation is dependent on them. Clearly, we should be reduced to intellectual, moral, and æsthetic chaos and nullity in a very short time. If a similar state of affairs should continue for some few generations, Britain would very soon cease to be of any importance in the world. The force of national traditions might keep up a certain unity; but we should be a people, or a crowd, living in the past, without energy, without pride in the present or hope in the future, having perhaps a little melancholy national sentiment, but incapable of national thought or action.

The continuance of the power and prosperity and unity of national life, the continued existence of the national mind and character, depends, then, upon the continued production of numbers of such men of more than average capacity. It is these men who keep alive from generation to generation, and spread among the masses and so render effective, the ideas and the moral influence of the men of supremely great powers. These men exert a guidance and a selection over the cultural elements which the mass of men absorb. They praise what they believe to be good, and decry what they believe to be bad; and, in virtue of the prestige which their exceptional powers have brought them, their verdict is accepted and moulds popular opinion and sentiment.

Consider how great in this way has been the influence of men like Carlyle, Matthew Arnold, and Ruskin. The tone and standard of taste, thought, and sentiment are set and maintained by such men. It is in their minds chiefly that the system of ideals and sentiments, which are the guiding principles and moving forces of the national mind, is perpetuated. They are truly "the salt of the earth"; without them the nation would soon fall into fragments, or become an inert and powerless mass of but low degree of organisation and unity.

It is because the national ideals and sentiments are formed by these leading spirits, and are perpetuated and developed by them, and by them impressed in some degree upon the mass of the people, and because in all national movements their influence predominates, that the judgments and actions, the character and the sentiments, of a nation may be different from, and in the higher nations are superior to, those of the average men of the nation. As Fouillée said—"The national character is not always best expressed by the mass, by the vulgar, nor even by the actual majority. There exists a natural *élite* which, better than all the rest, represents the soul of the entire people, its radical ideas, and its most essential tendencies. This is what the politicians too often forget."[1] That is to say, it is what they forget when, as is too often the case in this country, they consider that no movement must be undertaken till the mass of the people demands it. They ignore the fact that leadership is essential to the maintenance of national life at a high level, and, instead of exercising

[1] *Psychologie du peuple français*, p. 13.

initiative, they wait for it to come from below—wait for a mandate, as they say. The late President Roosevelt was a fine example of the contrary type of statesmanship. The character of the talents displayed by these exceptionally gifted individuals determines largely the form of the civilisation and, through shaping the social environment, tends to bring the minds of the mass of individuals more or less into harmony with it, giving them something of the same tendencies.

The men of genius of certain peoples, more especially peoples of relative racial purity, have excelled in some one direction. Thus the Semites have produced great religious teachers and little else, and have given to the world its three great monotheistic religions. The Tartar race has produced from time to time great soldiers and little else. It has made immense conquests and established dynasties ruling over other peoples. But, as in the case of the Turks who owe their national existence to a line of great despots of the house of Othman, they make little progress in civilisation and they do not unify the peoples they rule; for they produce ability of no other kind.

We see in most of the leading European nations the predominance of certain forms of genius. Modern Italy boasts chiefly men great in religion and art, perhaps owing to the predominance of *Homo Mediterraneus;* Spain in pictorial art and military conquest; England in poetry and administration and science; Germany in music and philosophy. Nevertheless, each of these peoples has produced men of the greatest power in all or several kinds; and this we may connect with the fact that they are all of very mixed racial composition. And we may add that France, the most composite or mixed racially, has produced the greatest variety of genius.

The production of the largest numbers of eminent men by peoples of mixed and blended racial elements, not too widely different, is what biological knowledge would lead us to expect. For, if a subrace is produced by crossing of varieties, it will be one of much greater variability than a pure race; as we see in the cases of the domesticated horse and dog and pigeon, of which the modern varieties are only kept pure by continual rejection of the departures from the standards, and of which the great variability renders possible the production, by selection, of very marked new features in a brief period of time.

The many elements which go to form the mental constitution of an individual become, in a mixed race, variously combined. If the crossed races are very widely different, the results seem to be in nearly all cases bad. The character of the cross-bred is made up of divergent inharmonious tendencies, which give rise to internal conflict, just as the physical features appear in bizarre combination; what examples we have—the Spanish Americans, the Eurasians, the Mulattoes, the half-breeds of Java and Canada—seem to show that a people so composed will produce few great men and will not become a great nation.

But, when the crossed races are less widely divergent, the elements of which the mental constitution is composed (and which direct observation and analogy with physical heredity show to be transmitted more or less independently of one another from parent to offspring) have opportunities to come together in new combinations, which result in mental constitutions unlike those of either parent (that is to say, the cross-breds are variable); and among these new combinations, while some will form minds below the average, others will form minds above the average in various degrees; and these, so long as the constitution is not too much weakened by radical lack of harmony of its elements, will be the effective great men.

Incidentally, these considerations perhaps throw light on a fact much discussed—namely, that exceptional powers, especially when of highly specialised nature, are often exhibited by persons of unstable mental constitution; whence arises the popular belief that genius is allied to, or is a form of, insanity.

These considerations also raise a presumption that peoples derived by the blending of several stocks may be expected to have progressed further in civilisation and in national growth than those of purer stock; and that, while the racial purity of a people may give stability, such a people will be liable to arrest and crystallisation of civilisation at an early stage, before culture is sufficiently advanced to render possible a highly developed national life. These indications are well borne out by a survey of the peoples of the world. We may see here, in all probability, one of the main causes of the early crystallisation of Chinese civilisation. Homogeneity and racial purity have produced extreme stability, but at the cost of the variability which produces great and original minds and, therefore, at the cost of capacity for national progress beyond an early stage.

CHAPTER X

Other Conditions of National Life

IN the two foregoing chapters, we have considered in relation to the life of nations three principal conditions essential to all collective mental life and action, even that of the unorganised crowd—namely, homogeneity, free communications, and leadership. We have now to consider other conditions which may render the collective mental processes of nations very different from, and superior to, those of a mere crowd.

In considering a patriot army as exemplifying collective life of a relatively high level, we distinguished five principal conditions that raise it above the level of the mental life of the crowd, in addition to one which is present in some crowds. This last was a common well-defined purpose present to, and dominant in, the minds of all individuals. It is this condition mainly that renders the collective mental life of such an army so simple, so relatively easy to understand, and so extremely effective.

This condition—a clearly defined common purpose dominant in the minds of the great mass of the constituent units—is for the most part lacking in the life of nations; its absence is one of the principal reasons for the ineffectiveness and bewildering complexity of their mental life. It is, however, occasionally realised in national life, and then we see how immense is its influence. Such an occasion is a war for national existence. Consider how, when the excesses of the French Revolution excited all the monarchies of Europe to attack France, the French nation, becoming animated with the one strong purpose of asserting its right to exist and to choose its own form of government, successfully drove back all its enemies and rose to a height of power and glory greater than at any other period; and how at the same time, its parts were welded more firmly together, so that it displayed a high degree of unity as well as of efficiency. Having achieved this high degree of unity and efficiency, the French nation, led on by the ambitions of Napoleon, became aggressive. And we are told by the historians that the attacks of Napoleon upon the various European peoples, which threatened to destroy whatever degree of national life those peoples had attained, were like the blows of a smith's hammer and resulted in welding together and hardening into nations the loosely aggregated races ruled over by the various monarchs; and that in this way these attacks initiated the modern period of Nation-States.[1]

War for national existence unifies nations. So long as the nation is not utterly shattered and crushed, such war greatly develops the national mind; because it makes one common purpose dominate the minds of all the citizens.

We are told that it is a practical maxim of cynical rulers to plunge their people into war when they are faced by dangerous internal discontents; and the reason usually given is that war diverts the attention of the people from their domestic grievances. But if it is a national war, a war in which the national existence is at stake, it does far more than merely divert attention; it binds the nation into a harmonious efficient whole by creating a common purpose; whereas, if the war is not of this order and is waged in some distant country and merely for some territorial aggrandisement, it has little or no such effect. Thus the

[1] Ramsay Muir, *op. cit.*, and J. Holland Rose, *The Development of the European Nations*, 1905.

recent Russo-Japanese war did little or nothing at the time to raise the Russian people in the scale of nationhood; it was followed by a period of national weakness; the national existence was not endangered, the objects of the war were too remote from the interests of the mass of the people to appeal to them strongly. Whereas the same war and the years of preparation for it, following upon the previous Chino-Japanese war, have made the Japanese one of the most efficient and harmonious nations of the world.

Another striking example of the same principle was the formation of modern Bulgaria as a strong Nation-State out of a population of quiet peasant proprietors united only by spatial proximity and by their racial distinctness from the surrounding populations. This creation of a strong nation out of a mere population of peasants was in the main the work of the war of 1885, by which the unprovoked attack of Servia was triumphantly repelled.[1]

The unity and nationhood of modern Germany is largely due to similar causes; and the war of 1871 may fairly be said to have led to a further integration of the national life of the French people, in spite of their defeat. America owes something of the same kind to the Spanish war; and the entry of that nation into the Great War, long delayed as it was, will probably be found to have had a similar effect. The French and Italian nations have undoubtedly been welded more firmly by the Great War; while England and her sister and daughter nations (with the one sad exception of the Irish) have been united, by their cooperation in the one great purpose, to a degree which no other conceivable event could have achieved and which many generations of peaceful industry and enlightened political efforts might have failed to approach.

History offers no parallel to these effects of war; and it

[1] J. Holland Rose, *op. cit.*

is difficult or impossible to imagine any other common purpose which could exert this binding influence in a similar degree. But it is worth while to notice that other and minor forms of international rivalry have corresponding effects. The international rivalry in aeronautics affords a contemporary illustration. Perhaps every one in this country has felt some degree of interest and satisfaction in the achievements of the adventurous spirits of our nation who have traversed the Atlantic by air. And it is probably largely owing to the prevalence of this national pride and purpose that, at a time demanding strict national economy, no voice has been raised against the enormous current expenditure of the government upon aeronautics.

Another and more important effect of the same kind is produced by the assumption of great national responsibilities in the way of administration in respect of backward peoples and undeveloped territories. The greatest example in history is the responsibility of Great Britain for the administration of India, gradually and only half-consciously assumed, but now keenly felt as at once a legitimate ground of national pride and a moral responsibility that cannot be laid aside. It is like the responsibility of the father of a family in its semi-instinctive origin and in its effects in steadying and strengthening character, for it imposes a responsibility which the nation, like the individual, cannot discharge indifferently without seriously damaging its reputation and prestige in the eyes of the world. Holland owes some of the strength of her nationhood to such influences; and the assumption by the American nation of responsibility for the peoples of Cuba and of the Philippine islands cannot fail to bring them in some degree similar moral benefits.[1]

[1] I suggest that international emulation in this sphere may prove to be an effective, probably the only effective, substitute for war.

Of the five other conditions of the higher development of a collective mind, let us notice, first and very briefly, continuity of existence, material and formal. Of course every nation has this in some degree, but some have it in much higher degree than others. The English nation is fortunate in this respect also. It has preserved both its formal and its material continuity in very high degree throughout many centuries, in fact ever since the Norman Conquest. No European nation can compare with it in this respect; it is only surpassed by China and perhaps Japan. The French nation has preserved its material continuity, its population and territory, in high degree. But the Great Revolution cut across and destroyed to a great extent its formal continuity, so that, as is sometimes said, the French nation has cut itself off from its past and made a new start; although, in doing so, it did not get rid of its highly centralised system of administration. The modern Italian and German nations are quite recent growths, their formal continuity having been subject to many interruptions. Spain, with her almost insular position, might have had continuity; but it was greatly disturbed by the imperial ambitions of her rulers in the sixteenth century and by the expulsion of the Moors. Greece is a striking example of loss of both material and formal continuity. The population of ancient Greece, which put her in the van of civilisation, has been largely abolished and supplanted by a different race; and her formal continuity also has suffered a number of complete ruptures.

Now material and formal continuity is, as we said, the essential presupposition of all the other main conditions of development of the collective mind. On it depends the strength of custom and tradition and, to a very great extent, the strength of national sentiment. It is, therefore, a principal condition of national stability; from it arise all the great conservative tendencies of the nation, all the forces that resist change; accordingly, the more complete and long enduring such continuity has been in the past, the greater is the prospect of its prolongation in the future. It is owing to the unbroken continuity of the English nation through so long a period that its organisation is so stable, its unwritten constitution so effective, at once stable and plastic, its national sentiment so strong, its complex uncoded system of judge-made law so nearly in harmony with popular feeling and therefore so respected. National organisation resting upon this basis of custom and traditional sentiment is the only kind that is really stable, that is not liable to be suddenly overthrown by internal upheavals or impacts from without. For it alone is rooted in the minds of all citizens in the forms of habit and sentiment. All other organisation is imposed by authority.

In this respect modern England and Germany offer a striking contrast that forces itself upon the most casual observation. As regards the mass of the people, the position of each individual in the organism of the German nation is officially determined by the written and codified law of the State; all personal status and relations are formally determined by official positions in this recently created system. Almost every individual carries about some badge or uniform indicating his position within the system. In England, the status and relations of individuals are determined by factors a thousand times more subtle and complex, involving many vaguely conceived and undefined traditions and sentiments. In Germany, it is almost true to say, if a man has no official position he has no position at all. In England, the comparatively few persons who have official positions have also their social positions by which their private relations are determined. They are officials only in their offices; whereas the German official is an official everywhere.

Other important topics we have to consider are (1) the organisation of the national mind; (2) the national self-consciousness; (3) the interaction of the nation as a whole with other nations. All these we may advantageously consider in the light of an analogy, the analogy between the individual mind and the collective mind of the nation. This is a much closer and more illuminating analogy than that between the nation, or society, and the material organism. The latter analogy has been developed in detail by H. Spencer, Schäffle,[1] and others; it has now fallen into some disrepute. It has no doubt a certain value, but it is popularly used in a way that leads to quite unjustifiable conclusions. Of these fallacies by far the most commonly accepted is that which asserts that, just as every animal organism inevitably grows old and dies, so too must nations.

This is one of the most popular dogmas of amateur philosophers, and so distinguished a statesman as the late Lord Salisbury gave it countenance; while Mr. A. J. Balfour in his recent Sidgwick Memorial Lecture[2] courageously breaks away and proposes to substitute for senility as the cause of decay the word *decadence*—a proposal which merely implies that he trusts less to the analogical argument from the material organism and more to empirical induction, to the observation of the fact that so many nations have decayed.

All this serves to illustrate the dangers of analogy. We need no special cause to account for the fall and the decay of nations, no obscure principle of senility or decadence; the wonderful thing is that they exist at all; and what needs explanation is not so much the decay of some, but rather the long persistence of others.

Let us turn, then, to the analogy between the organisation of the national collective mind and that of the in-

[1] *Bau und Leben des socialen Körpers.* [2] *Decadence.*

dividual mind, which, I say, is so much closer and more illuminating than that between a society and a bodily organisation.

The actions of the individual organism are the expression of its mental constitution or organisation; in some creatures this organisation is almost wholly innate—the organisation consists of a number of reflex and instinctive dispositions each specialised for bringing about a special kind of behaviour under certain circumstances. Such old established racial dispositions with their special tendencies have their place in more complexly developed minds; but in these their operations are complicated and modified by the life of ideas, and by a variety of habits developed under the guidance of ideas and in the light of individual experience.

The enduring reflex and instinctive dispositions of the individual mind we may liken to the established institutions of a nation, such as the army and navy, the post office, the judicial and the administrative systems of officials. These, like the instincts, are specialised executive organisations working in relative independence of one another, each discharging some specialised function adapted to satisfy some constantly recurring need of the whole organism. In both cases such semi-independent organisations, the instincts or the institutions, are relatively fixed and stable, and they work, if left to themselves, quasi-mechanically along old established lines, without intelligent adaptation to new circumstances; and they are incapable of self-adaptation. In both cases, the mental organisation is in part materialised, the instinct in the form of specialised nervous structure, the institution in the form of the material organisation essential to its efficient action, the buildings, the printed codes, the whole material apparatus of complex national administration. In both cases, the actions in which they play their part are not purely mechanical but to some extent truly psychical—though of a low order.

If we accept the view, which is held by many, that instincts and reflexes are the semi-mechanised results of successive mental adaptations effected by the mental efforts of successive generations, then the analogy is still closer; for the permanent national institutions are also the accumulated semi-mechanised products of the efforts at adaptation of many generations.

The organisation of some nations resembles that of the minds of those animals whose behaviour is purely instinctive. Such is a nation whose organisation takes the form of a rigid caste system. Each caste performs its special functions in the prescribed manner in relative independence of all the others. And, in both cases, the organisation of the mind includes no means of bringing the different fixed tendencies or dispositions into harmonious co-operation in the face of unusual circumstances. The whole system lacks plasticity and adaptability; for it is relatively mechanical and of a low degree of integration. Any true adaptation of the whole organism by mental effort is impossible in both cases.

The higher type of individual mind is characterised by the development of the intellectual organisation by means of which the activities of the various instincts, the executive organisations, may be brought into co-operation with, or duly subordinated to, one another; and the activities of each such individual may be further adapted to meet novel combinations of circumstances not provided for in the innate organisation; hence, the activities of the whole organism, instead of being a succession of quasi-mechanical actions, and of crude conflicts between the impulses or tendencies of the different instincts, reveal a higher degree of harmony of the parts, a greater integration of the whole system, and a much greater adaptability to novel circumstances; while, at the same time, the behaviour of the whole, in face of any one of the situations provided for by innate organisation or instinct, is liable to be less sure and perfect than in the case of the less complex, less highly evolved type of mind.

Exactly the same is true of the more highly evolved type of national mind. Like the lower type, it has its executive institutions and hierarchies of officials, organised for the carrying out of specialised tasks subserving the economy of the whole. But, in addition, it has a deliberative organisation which renders possible a play of ideas; and, through this, the operations of the institutions are modified and controlled in detail and are harmonised in a way which constitutes a higher integration of the whole.

In both cases ideas and judgments reached by the deliberative processes can only become effective in the world of things and conduct by setting to work, or calling into play, one or more of the executive dispositions or institutions.

In both cases, ideas and the deliberative processes, which to some extent control the operations of the innate or traditional dispositions, produce, in so doing, some permanent modification of them in the direction of adaptation to deal with novel circumstances; so that the dispositions or institutions grow and change under the guidance of the deliberative processes, slowly becoming better adapted for the expression of the ruling ideas; they become better instruments, and more completely at the service of ideas and of the will.

Just as the animal, on the instinctive plane of mental life, displays a very efficient activity in the special situation which brings some one instinct into play, so any one caste of a caste-nation may perform its function under normal circumstances with great efficiency, the priestly caste its priestly function, the warrior caste, or the caste of sweepers, its function; and, in both cases, the development of the deliberative organisation is apt to interfere to some extent with the perfect execution of these specialised functions.

Again, in the individual mind, adaptation of conduct to novel circumstances, or to secure improved action in familiar circumstances, requires the direction of the attention, that is the concentration of the whole energy of the mind, upon the task; whereas, when the new mode of behaviour is often repeated, it becomes more and more automatic; for, owing to the formation of new nervous organisation, the attention is set free for other tasks of adaptation. Just in the same way new modes of national behaviour are only effected when the attention of the nation's mind is turned upon the situation; whereas, with recurrence of the need for any such novel mode of action, there is formed some special executive organisation, say a Colonial Office, or an Unemployed Central Committee, or an Imperial Conference, which deals with it in a more or less routine fashion, and which, as it becomes perfected, needs less and less to be controlled and guided by national attention and therefore operates in the margin of the field of consciousness of the national mind, while public attention is set free to turn itself to other tasks of national adaptation.

We may also regard the customs of a nation as analogous with the habits of the individual, if (for the sake of the analogy) we accept the view that instincts are habits that have become hereditary; for custom is an informal mode in which routine behaviour is determined, and it tends to lead on to, and to become embodied in, formal institutions; it is like habit, a transition stage between new adaptation and perfected organisation. Individual adaptation, habit and instinct are parallel to national adaptation, custom, and legal institution.

At the risk of wearying the reader, I will refer to one last point of the analogy. Individual minds become more completely integrated in proportion as they achieve a full self-consciousness, in proportion as the idea of the self becomes rich in content and the nucleus of a strong sentiment generating impulses that control and override impulses of all other sources. In a similar way, the national mind becomes more completely integrated in proportion as it achieves full self-consciousness, that is, in proportion as the idea of the nation becomes widely diffused among the individual minds, becomes rich in content and the nucleus of a strong sentiment that supplies motives capable of overriding and controlling all other motives.

Consider now in the light of this analogy the principal types of national organisation. The organisation of some peoples is wholly the product of the conflicts of blind impulses and purely individual volitions working through long ages. This is true of many peoples that have not arrived at a national self-consciousness or, as the French say, a social consciousness, and are not held in servitude by a despotic power. It is a natural stage of evolution which corresponds to the stage of the higher mammals in the scale of evolution of the individual mind. A nation of this sort has no capacity for collective deliberation and volitional action. What collective mental life it has is on the plane of impulse and unregulated desire. Such ideas as are widely accepted may determine collective action; but such action is not the result of the weighing of ideas in the light of self-consciousness; hence they are little adapted to promote the welfare of the nation, and, because there is no organisation adapted for their expression, they can be but imperfectly realised.

We may perhaps take China (as she was until recently) as the highest type of a nation of this sort. Hers was a complex and vast organisation consisting of very ancient institutions and customs, slowly evolved by the conflict of impulses and in part imposed by despotic power and individual wills; not formed by a national will under the guidance of national self-consciousness.[1] Hence China was incapable of vigorous national thought or volition, and its nearest approach to collective action was expressed in such blind impulsive actions as the Taeping Rebellion or the Boxer Rising. This last seems to have been prompted by a dawning national self-consciousness which had not, however, so moulded the national organisation as to make it an efficient instrument of its will.

Of other nations the organisation is, in part only, a natural growth, having been, in large part, impressed upon it by an external power. Such is the case in all those many instances in which a foreign power of higher social organisation has conquered and successfully governed for a long period a people of lower civilisation. We may see a parallel to this type in the mind of an individual whose behaviour is in the main the expression of a number of habits engendered by a severe discipline which has continued from his earliest years, and which has never permitted the free development of his natural tendencies and character. Such was England under the feudal system imposed upon her by her Norman conquerors. Such also France under Louis XIV. Such was Russia when the Varegs, the conquering Northmen, imposed on the almost unorganised mass of Slavs their rule and a national organisation; and such it remained up to the outbreak of the Great War, a mass of men in whom the national consciousness was only just beginning to glimmer here and there, crudely organised by the bureaucratic power of a few. Even in the minds of these few the national consciousness and purpose was but little developed. Individual pur-

[1] Cf. A. Smith's *Village Life in China*. The author insists on the lack of public spirit, of the idea of action *pro bono publico*.

poses and individual self-consciousness predominated. Hence Russia had no capacity for national thought and action; and when, as recently, ideas stirred the masses to action, their actions were those of unorganised crowds, impulsive and ineffective; the ends were but vaguely conceived, the means were not deliberately chosen, or, if so chosen, found no executive organisation for the effective expression of the collective purpose.

In such nations the organisation, which has been in the main created by a small governing class, is adapted only for the execution of its purposes, and not at all for the formation of a national mind and the expression of the collective will. The organisation consists primarily in a system for the collecting of taxes and the compulsory service of a large army. The revenue is raised for two primary purposes—the support of the governing class or caste in luxury and the support of the army; and the end for which the army is maintained is primarily the gathering of the taxes, and the further extension of the tax-collecting system over larger areas and populations—a vicious circle. On the other hand, the conditions which tend to the formation of national mind and character (which would have quite other ends than these) are naturally suppressed as completely as possible by the governing few.

Russian history in modern times exemplifies these principles in the clearest and most complete manner. The effects of this sort of organisation were very clearly illustrated by Count Tolstoi's articles in the *Fortnightly Review*,[1] in which he expressed as his social creed and ideal a complete anarchy to be achieved by passive resistance; denied that nations have or can have any existence; and asserted that the idea of a nation is as fictitious as it is pernicious. He had in mind only this type of organisation of a people, which hardly entitles it to be called a nation.

[1] 1906.

And the same considerations explain the wide prevalence of philosophic anarchy in Russia.

Another type of national organisation results when the natural evolution of the national mind and character has been artificially and unhealthily forced by the pressure of the external environment of a people, when the need of national self-preservation and self-assertion compels the mass of the people to submit to an organisation which is neither the product of a natural evolution through the conflict of individual wills, nor the expression of the general mind and will, nor is altogether imposed upon it for the individual purposes of the few, but is a system planned by the few for the good of the whole, and by them imposed upon the whole. This is the kind of organisation of which a modern army stands as the extreme type and which is best represented among modern nations by Germany as she was before the War.

Under such a system there appears inevitably a tendency rigorously to subordinate the welfare of individuals to that of the nation as a whole. And that was just the state of affairs in Germany. German political philosophy showed the opposite extreme from Tolstoi's; the individual existed for the nation only. Hence we find this condition of affairs justified by such writers as Bluntschli,[1] Treitschke and Bernhardi, who represent the State as having an existence and a system of rights superior to that of all individuals; and we see attempts to justify the subordination of individual interests by means of the doctrine of the "collective consciousness."[2]

In such States as that of the foregoing type the one kind of organisation is alone highly developed, namely the executive organisation; while the deliberative organisation is very imperfect and is repressed and discouraged by the

[1] In his *Theory of the State*.

[2] By Schaeffle, *op. cit.*, and all the school of German "idealism."

governing power. Such a State is likely to appear very strong in all its relations with other States, and its material organisation may be developed in an effective and rapid way, as we have seen in pre-war Germany. But its actions are not the expression of the national will and are not the outcome of the general mind. They are designed by the minds of the few for the good not of all, but of the whole, the good, that is, not of individuals but of the State.

Organisation of this type is not of high stability, in spite of its appearance of strength and its efficiency for certain limited purposes, such as industrial organisation and the promotion and diffusion of material well-being. In a State so organised there inevitably grows up an antagonism between individual rights and interests and the rights and interests of the State. It is psychologically unsound. This fact was revealed in Germany by the tremendous growth of social democracy, which was the protest against the subordination of individual welfare to that of the State. The defect of such organisation was illustrated by the fact that Germany, though its well-governed population increased rapidly, for many years continued to lose great numbers of its population to other countries. For the mass of the people felt itself to be not so much of the State as under it. And it is, I think, obvious that the advent of a bad and stupid monarch might easily have brought on a revolution at any time.

The inherent weakness of the system induced the governing power to all sorts of extreme measures directed to maintain its equilibrium and cohesion. Among such State actions the gravest were perhaps the deliberate falsification of history by the servile historians and the suppression and distortion of news by the press at the command and desire of the State. The expropriation of the Polish landowners and the treatment of Alsace-Lorraine were other striking manifestations of the imperfect development of the national mind and of the corresponding practice and philosophy of the State-craft which the world has learnt to describe as Prussian.

The organisation of pre-war Germany was, then, very similar to that of an army and was efficient in a similar way, that is to say for the attainment of particular immediate ends. In a wider view, such national organisation is of a lower nature than that of England or France or America; for the ends or purposes of a nation are remote, they transcend the vision of the present and cannot be defined in terms of material prosperity or military power; and only the development of the national mind, as a natural and spontaneous growth, can give a prospect of continued progress towards those indefinable ends. Germany was organised from above for the attainment of a particular end, namely material prosperity and power among the peoples of the world; and, as the bulk of her population had been led to accept this narrow national purpose, the organisation of the nation, like that of an army, was extremely effective for the purpose. It gave her a great advantage as against the other nations, among whom the lack of any such clear-cut purpose in the minds of all was a principal difficulty in the way of effective national thought and action. For a like reason the existence of a nation organised in this way is a constant threat to the nations of higher type; and, as we have seen, it may compel them at any time to revert to or adopt, temporarily at least and so far as they are able, an organisation of the lower and more immediately effective kind. And this threat was the justification of the nations of the Entente, when they demanded a radical change in this political organisation of Germany. In a similar way, in the past, the Huns, the Turks, and the Arabs, peoples organised primarily for war and conquest, had to be destroyed as nations if the evolution of nations of higher type was to go forward.

CHAPTER XI

The Will of the Nation[1]

ROUSSEAU, in his famous treatise, *Le Contrat Social*, wrote: "There is often a great difference between the will of all and the general will; the latter looks only to the common interest; the former looks to private interest, and is nothing but a sum of individual wills; but take away from these same wills the plus and minus that cancel one another and there remains, as the sum of the differences, the general will." "Sovereignty is only the exercise of the general will." That is to say, a certain number of men will the general good, while most men will only their private good; the latter neutralise one another, while the former co-operate to form an effective force.

Dr. Bosanquet,[2] criticising Rousseau's doctrine, says that the general will is expressed by the working of the institutions of the community which embody its dominant ideas; that no one man really grasps the nature and relations of the whole society and its tendencies; that the general will is thus unconscious (by which he seems to mean that the nation is unconscious of itself and of its ends or purposes); and he goes on to say that the general will is the product of practical activities making for nearer smaller ends, and

[1] The substance of this chapter was contained in a paper entitled "The Will of the People," read before the Sociological Society and published in the *Sociological Review*, 1912.

[2] *Philosophical Theory of the State* and Article in *International Journal of Ethics*, 1907.

that its harmony depends on the fact that the activities of each individual are parts of a systematic whole.

Bosanquet's theory amounts to a justification of the old individualist *laissez faire* doctrine—the doctrine that the good of the whole is best achieved by giving freest possible scope to the play and conflict of individual purposes and strivings—the philosophic radicalism of Bentham and Mill, which teaches that, if each man honestly and efficiently pursues his private ends, the welfare of the State somehow results. How this systematic whole which is the State arises he does not explain; and it seems to me that Bosanquet leaves unsolved that difficulty which, as we saw, led Schiller and others to postulate an external power guiding each people—the difficulty, if we assume that individual wills strive only after private egoistic ends, of explaining how the good of the whole is nevertheless achieved.

In all societies many general changes result, and in some nations no doubt the good of the whole is achieved in a measure by fortunate accident, in the way Bosanquet describes—namely, by the interplay of individual wills working for near individual ends. But, I think, it is improper to say that in such a case any general will exists. Such a nation, if it displays any collective activity, only does so in an impulsive blind way which is not true volition, but is comparable rather with instinctive action; for, as we have seen, self-consciousness is essential to volition; a truly volitional action is one which issues from the contemplation of some end represented in relation to the idea of the self and found to be desirable. And the changes of a society which result in the way Bosanquet claims as the expression of the general will are unforeseen and unwilled; they are no more the expression and effects of a general will than are the movements of a billiard ball struck simultaneously by two or more men each of whom aims at a different position.

Bosanquet maintains that the national will is unconscious of its ends; but that the life of a nation does express a general will, in virtue of the fact that the individuals, who will private and less general ends than the ends of the nation, live in a system of relations that constitutes them an organism; and that it is in virtue of this organic system of relations that the individual volitions work out to an unforeseen unpurposed resultant, which he calls the end willed by the general will. He makes of this organic unity the essential difference between a mere crowd and a society or nation—defining an organism or organic unity as a system of parts the capacities and functions of each of which are determined by the general nature and principle of the whole group.

That is an excellent definition of an organism; and we have recognised fully the importance of organisation in national life, consisting in specialisation of the parts such that each part is adapted to perform some one function that subserves the life of the whole, while itself dependent upon the proper functioning of all other parts.

We may admit too that, in proportion as this specialisation of functions is carried further, organic unity is promoted because the life of the whole becomes more intimately dependent on the life of each part, and each part more intimately and completely dependent on the life of the whole.

But unity of this sort is characteristic of all animal bodies; and, though the mind has this kind of organic unity, it acquires, in proportion as self-consciousness develops, over and above this kind of unity, a unity of an altogether new and unique kind; a unity which consists in the whole (or the self) being present to consciousness, whether clearly or obscurely, during almost every moment of thought, and pervading and playing some part in the determination of the course of thought and action.

Now, the national mind also has both the lower and the higher kinds of unity. In both cases—that is, in the development both of the individual and of the national mind—a certain degree of organic unity must be achieved, before self-consciousness can develop and begin to play its part; but in both cases, when once it has begun to operate, self-consciousness goes on greatly to increase the organic unity, to increase the specialisation of functions and the systematic interdependence of the parts.

Consider a single illustration of this parallelism of the individual with the national mind. Take the æsthetic faculty. In the individual mind there develops a certain capacity for finding pleasure in certain objects and impressions, such as young children and even the animals have; and then, with the growth of self-consciousness, the individual sets himself deliberately to cultivate this faculty, to specialise it along particular lines and to exercise it as something apart from his other mental functions; while, nevertheless, it becomes for him, in proportion as it is developed and specialised, more and more an essential part of his total experience. Just so there spontaneously develop in a people some rudimentary æsthetic practices and traditions and some class of persons, say the bards, who are more skilled than other men in ministering to the æsthetic demands of their fellows. Then, as national self-consciousness develops, the place and value of these functions in the system of national life becomes explicitly recognised, and they are deliberately fostered by the establishment of national institutions, schools of art, academies of letters and music, the award of public titles and honours and so forth; whereby the specialisation of these national functions is increased, their dependence on the life of the whole rendered more intimate, and, at the same time, the life of the whole rendered more dependent upon the life of these parts, because the richer æsthetic development of the parts reacts upon the whole, diffusing itself through and elevating the life of the whole.

Bosanquet recognises in national life only the lower kind of unity and not the unity of self-consciousness. He seems to reject the notion of national self-consciousness, on the ground that the life of a nation is so complex that it cannot be fully and adequately reflected in the consciousness of any individual; yet in this respect the difference between the national mind and the individual mind is one of degree only and not of kind. In the individual mind also, even the most highly developed and self-conscious, the capacities and dispositions and tendencies that make up the whole mind are never fully and adequately present to consciousness; the individual never knows himself exhaustively, though he may continually progress towards a more nearly complete self-knowledge. Just so the national mind may progress towards a more complete self-knowledge, and, though at the present time no nation has attained more than a very imperfect self-knowledge, yet the process is accelerating rapidly among the more advanced nations; and such increasing self-knowledge promises to become the dominating factor in the life of nations, as it is in the lives of all men, save the most primitive.

Suppose that all those conditions making for national unity which we have considered in the foregoing chapters were realised, but that nevertheless all men continued to be moved only by self-regarding motives, or by those which have reference to the welfare of themselves and their family circle or to any ends less comprehensive than the welfare of the whole nation. We could not then properly speak of the tendencies resulting from the interplay and the conflict of all these individual wills as expressions of the general will, as Bosanquet and others have done, even though the organic unity of the whole secured a harmonious resultant national activity, if such a thing were possible. But there is no reason to suppose that such a thing is possible.

I think we may say that it is only in so far as the idea of the people or nation as a whole is present to the consciousness of individuals and determines their actions that a nation in the proper sense of the word can exist or ever has existed. Without this factor any population inhabiting a given territory remains either a mere horde or a population of slaves under a despotism. Neither can be called a nation; wherever a nation has appeared in the history of the world, the consciousness of itself as a nation has been an essential condition of its existence and still more of its progress.

We may see this, even more clearly, in the case of the smaller aggregations of men, the smaller social units, the family, the clan, the tribe. The family is a family only so long as it is conscious of itself as a family, and only in virtue of that self-consciousness and of the part which this idea and the sentiments gathered about it play in determining the actions of each member. How carefully such family consciousness is sometimes fostered and how great a part it plays in social life is common knowledge. In the early stages of Greek and Roman history, the family consciousness was the dominant social force which long succeeded in overriding and preventing the development of any larger social consciousness. Just as the *gens* played this part in early Rome, so the clan has played a similar part elsewhere, for example in the highlands of Scotland. Such peoples form the strongest nations.

Just so with the tribe. It exists as a tribe only because, and in so far as, it is conscious of itself, and in so far as the idea of the tribe and devotion to its service determines the actions of individuals. The mere fact of the possession of a tribal name suffices to prove the existence of this self-consciousness. And, as a matter of fact, tribal self-consciousness is in many cases extremely strongly developed; the idea of the tribe, of its rights and powers, of its past and its future plays a great part among warlike savages; and an injury done to the tribe, or an insult offered to it, will often be kept in mind for many years, even for generations, and will be avenged when an opportunity occurs, even in spite of the certainty of death to many individuals and the risk of extermination of the whole tribe.

The federation of Iroquois tribes to form a rudimentary nation seems to have been due to a self-conscious collective purpose. And, when other tribes become fused to form nations, the same holds true. Consider the Hebrew nation, one of the earliest historical examples of a number of allied tribes becoming fused to a nation. Surely the idea of the nation as the chosen people of Jehovah played a vital part in its consolidation, implanted and fostered as it was by a succession of great teachers, the prophets. Their work was to implant this idea and this sentiment strongly in the minds of the people, to create and foster this traditional sentiment by the aid of supernatural sanctions. The national self-consciousness thus formed has continued to be not only one factor, but almost the only factor or condition, of the continued existence of the Jewish people as a people, or at any rate the one fundamental condition on which all the others are founded—their exclusive religion, their objection to intermarriage with outsiders, their hope of a future restoration of the fortunes of the nation, and so forth.

And the same is true of every real nation; its existence and its power are grounded in its consciousness of itself, the idea of the nation as a dominant factor in the minds of the individuals. The dominant sentiment which centres about that idea is very different in the various nations.

It may be chiefly pride in the nation's past history, as in Spain; or hope for its future, as in Japan; or the need of self-assertion in the present, as in pre-war Germany.

The political history of Europe in the nineteenth century is chiefly the history of the national actions that have sprung from increase of national self-consciousness resulting from the spread of education, from the improvement of means of communication within each people and from increase of intercourse between nations. The opening pages were the wars in which the French people, suddenly aroused to an intense national consciousness, successfully resisted and drove back all the other European powers. Of other leading events the formation of modern Italy and of modern Germany, of Bulgaria, Serbia, Greece, were results of the awakening of national self-consciousness.[1]

The resistance of the Japanese to the Russians and their victory over them were in the fullest sense the immediate outcome of the idea of the Japanese nation in the minds of all its people, leading to a strong collective volition for the greater power, glory, and advancement of the nation. The recent unrest in China is recognised on all hands to be the expression of a dawning national self-consciousness. In Europe, Poland, Finland, Hungary, and Ireland exemplify its workings very clearly in recent years. The Magyars were not oppressed by the Austrians. They, economically and individually, had nothing material to gain by a separation from Austria; and in separating themselves they would have risked much, their lives, and their material welfare; yet the idea of the Magyar nation impelled them to it. The Poles of Germany were not rebellious because they were ill-treated and their affairs maladministered. If they could and would have cast out from their minds the idea of the Polish nation, they might

[1] Cf. J. H. Rose, *The Development of the European Nations* and Ramsay Muir, *Nationalism and Internationalism.*

have comfortably shared in the marvellously advancing material prosperity of Germany. But they were severely treated by the Germans, because they were moved by this idea and this sentiment; and the bad treatment it brought upon them did but render the idea more vividly, more universally, present to the consciousness of all, even of the little children at school, and, by inflaming the passions which have their root in the national sentiment, strengthened that sentiment.

But for the idea of the Boer nation and the dawning national sentiment, the late Boer war would never have occurred; and that sentiment was, as in the case of the Japanese in their late war, the principal source of the great energy displayed by the Boers and of such success as they achieved.

Even in India, the proposal to divide Bengal has suddenly discovered among the Bengalese, the most submissive part of the population, the part which has seemed most devoid of national spirit, the existence and the importance as a political factor of the idea of the Bengalese as a people and of sentiments centred upon that idea.

The rapid increase of national self-consciousness among the peoples of the world and the increasing part everywhere played by the sentiment for national existence are in short the dominant facts in the present period of world history; their influence overshadows all others.

Since, then, any nation exists only in virtue of the existence of the idea of the nation in the minds of the individuals of whom it is composed, and in virtue of the influence of this idea upon their actions, and since this idea plays so great a part in shaping the history of the world, it is absurd to maintain that the general will is but the blind resultant of the conflict of individual wills striving after private ends and unconscious of the ends or purposes of the nation. In opposition to such a view, we must maintain that a population seeking only individual ends cannot form or continue to be a nation, though all the other conditions we have noticed be present; that a nation is real and vigorous in proportion as its consciousness of its self is full and clear. In fact national progress and power and success depend in chief part upon the fulness and the extension, the depth and width of this self-consciousness—the accuracy and fulness with which each individual mind reflects the whole; and upon the strength of the sentiments which are centred upon it and which lead men to act for the good of the whole, to postpone private to public ends. And the same holds good of all the many forms of corporate life within the nation. Each individual's sense of duty, in so far as it is a true sense of duty, and not a fictitious sense due merely to superstitious fear or to habit formed by suggestion and compulsion, is chiefly founded upon the consciousness of the society of which he forms a part, upon the group spirit that binds him to his fellows and makes him one with them. And the nations in which this national self-consciousness is strongest and most widely diffused will be the successful nations.

Reflect a little on these facts vouched for by General Sir Ian Hamilton. No soldier of the Japanese army, none even of the coolies, would accept anything in the shape of a tip even for honest services rendered, lest the purity of his motives should be sullied; and each man always went into action not merely prepared to die if necessary, but actually prepared and expecting to conquer and to die for the good and glory of his nation. He writes "Japanese officers have constantly to explain to their men that they must not consider the main object of the battle is to get killed."[1] And he goes on to show that they are not fanatics, are not inspired by any idea of the supernatural or by any hope of rewards after death, as is usually the case of the Moslem

[1] *The Russo-Japanese War*, Vol. II, p. 25.

soldier who displays an equal recklessness of life. Surely, if in any nation the national consciousness could inspire and maintain all classes of its people in all relations of life to this high level of strenuous self-sacrifice for the welfare of the nation, that nation would soon predominate over all others, and be impregnably strong, no matter what defects of individual and national character it might display.

The idea of the nation is, then, a bond between its members over and above all those bonds of custom, of habit, of economic interdependence, of law and of self-interest, of sympathy, of imitation, of collective emotion and thought, which inevitably arise among a homogeneous people occupying any defined area; and it is the most powerful and essential of them all. As Fouillée put it, the essential characteristic of human society is that "it is an organism which realises itself in conceiving and in willing its own existence. Any collection of men becomes a society in the only true sense of the word, when all the men conceive more or less clearly the type of organic whole which they can form by uniting themselves and when they effectively unite themselves under the determining influence of this conception. Society is then an organism which exists because it has been thought and willed, it is an organism born of an idea."[1] In this sense Society has never yet been perfectly realised, but it is the ideal towards which social evolution tends.

National group self-consciousness plays, then, an all-important part in the life of nations, is in fact the actual, the most essential constitutive factor of every nation; and nationhood or the principal of nationality is the dominant note of world history in the present epoch; that is to say, the desire and aspiration to achieve nationhood, or to strengthen and advance the life of the nation, is the most powerful motive underlying the collective actions of

[1] *La Science Sociale contemporaine*, p. 115.

almost all civilised, and even of semi-civilised, mankind; and the consequent rivalry between nations overshadows every other feature of modern world history, and is convulsing and threatening to destroy the whole of modern civilisation. It is surely well worthy of serious study. Yet, owing to the backward and neglected state of psychology, not only is this study neglected, but, as we have seen, some of our leading political philosophers have not yet even realised the essential nature of the problem; and many of the historians, economists, and political writers are even further from a grasp of its nature. They have been forced by the prominence and urgency of the facts to recognise what they call the principle of nationality; and even now the majority of them are demanding that, in the European settlement and in the affairs of the world in general, the principle of nationality shall be given the leading place and the decisive voice. But they do not recognise that the understanding of this principle, this all-powerful political factor, is primarily and purely a psychological problem. We find them, in discussing the nature of nations and the conditions of nationality, perhaps mentioning the psychological view of nations as a curious aberration of a few academic cranks, from which they turn to discover the true secret of nationality in such considerations as geographical boundaries, race, language, history, and above all economic factors; they do not see that each and all of these conditions, real and important though they are and have been in shaping the history and determining the existence of nations, only play their parts indirectly by affecting men's minds, their beliefs, opinions, and sentiments, especially by favouring or repressing the development in each people of the idea of the nation.

The all-dominant influence of the idea of the nation, I insist, is not a theory or a speculative suggestion, it is a literal and obvious fact. Let every other one of the favouring conditions of nationality, the geographical, historical, economic be realised by a population; yet, if that population has no collective self-consciousness, is not strongly actuated to collective volition by the group spirit, it will remain not a nation, but a mere aggregate of individuals, having more or less organic unity due to the differentiation and interdependence of its parts, but lacking that higher bond of unity which alone can ensure its stability and continuity, and which, especially, can alone enable it to withstand and survive the peaceful pressure or the warlike impact of true nations.

I am not at present defending nationality; I shall come back to the question of its value. I am now only concerned with the psychological problem of the nature and conditions of the development of national self-consciousness. I have been using the latter phrase and the phrase "the idea of the nation" as a shorthand expression; but I must remind the reader that we have to beware of the intellectualist error of regarding ideas as moving powers; ideas as merely intellectual representations or conceptions have no motive power, they are in themselves indifferent. It is only in so far as the object conceived becomes the object of some sentiment that the conception of it moves us strongly to feeling and action. I must refer, therefore, to what I have written on the sentiments and the self-regarding sentiment.[1] Here I would insist on the strictness in this point of our analogy between the individual and the national mind. I have pointed out that the

[1] *Social Psychology*, Chapters V—IX. Dr.Bosanquet's failure (as it seems to me) to achieve a satisfactory account of the social will is the inevitable consequence of the inadequacy of his conception of individual volition. This is set out in his *Psychology of the Moral Self*, where he shows himself to be an uncompromising adherent of the intellectualist tradition. He totally ignores the existence and organisation of the conative side of the mind. His notion of volition is based upon the now discredited theory of ideo-motor action.

individual's idea of himself only develops beyond a rudimentary stage because and in so far as this idea becomes the nucleus of a strong self-regarding sentiment which gives him an interest in himself, directs his attention upon his own personality and its relations, and impels him to strive to know himself. So that a developed individual self-consciousness never is and never can be a purely intellectual growth; it always involves a strong sentiment, a centring of emotional conative tendencies upon this object, the self. Exactly the same is true of nations.

Hence national self-consciousness can never develop except in the form of an idea of strong affective tone, that is to say a sentiment. Hence, whenever we speak of national self-consciousness or the idea of the nation as a powerful factor in its life, the sentiment is implied, and I have implied it when using these expressions hitherto. This national sentiment, which, if we use the word in its widest sense, may be called patriotism, is, like all the other group sentiments, developed by way of extension of the self-regarding sentiment of the individual to the group, and may be further complicated and strengthened by the inclusion of other tendencies. A point of especial importance is that this great group sentiment can hardly be developed otherwise than by way of extension of sentiments for smaller included groups, the family especially. For the idea of the nation is too difficult for the grasp of the child's mind; the nation cannot become the object of a sentiment until the intellectual powers are considerably developed. Hence the development of a family sentiment, or of one for some other small easily conceived group, is essential for the development in the child of those modes of mental action which are involved in all group feeling and action. For this reason the family is the surest, perhaps essential, foundation of national life; and national self-consciousness is strongest, where family life is strongest.

The development of the group spirit in general and of national self-consciousness in particular is favoured by, and indeed dependent upon, conditions similar to those which develop the self-consciousness of individuals. Here is another striking point of the analogy between the individual and the national mind. Passing over other conditions, let us notice one, the most important of all. The individual's consciousness of self is developed chiefly by intercourse with other individuals—by imitation, by conflict, by compulsion, and by co-operation. Without such intercourse it must remain rudimentary. The individual's conception of himself is perpetually extended by his increasing knowledge of other selves; and his knowledge of those other selves grows in the light of his knowledge of himself. There is perpetual reciprocal action. The same is true of peoples. A population living shut off, isolated from the rest of the world, within which no distinctions of tribe and race existed, would never become conscious of itself as one people and, therefore, would not become a nation. Some such conditions obtained for long ages among the pastoral hordes of the central Eurasian Steppes, which, so long as they remained there, have never formed a nation; and the same was true of the tribes of Arabia, until Mahomet impelled them by his religion of the sword to hurl themselves upon neighbouring peoples.

Of civilised peoples, China has had least intercourse with the outer world. The Chinese knew too little of other races to imitate them; they did not come into conflict or co-operation with others, save in a very partial manner at long intervals of time, or only with their Mongol conquerors, whom they despised as inferior to them in everything but warfare, and whom they abhorred. Hence, in spite of the homogeneity of the people, of the common culture, and of the vast influence of great teachers, national consciousness and the group spirit in all its forms remained at a low level. Hence, a great deficiency in those virtues which have their root in the social consciousness; a low standard of public duty, a lack of the sense of obligation to society. Hence, the corruptness and hollowness of all official transactions and political life. Want of honesty in public affairs is not the expression of an inherent defect of the Chinese character; for in commercial relations with Europeans the Chinaman has proved himself extremely trustworthy, much superior indeed in this respect to some other peoples. It is probable that, if China, like Europe, had long ago been divided into a number of nations, each of them, through action and reaction upon the rest, would have developed a much fuller national consciousness than exists at present and some considerable degree of public spirit, and would consequently have advanced very much farther in the scale of social evolution, instead of standing still, as the whole people has done for so long.

Everywhere we can see the illustrations of this law. Of all forms of intercourse, conflict and competition are the most effective in developing national consciousness and character, because they bring a common purpose to the minds of all individuals; and that is the condition of the highest degree and effectiveness of collective mental action and volition. It is under these conditions that the idea of the nation and the will to protect it and to forward its interests become predominant in the minds of individuals; and the more so the greater the public danger, the greater and the more obvious the need for the postponement of private ends to the general end.

Already there is beginning to develop a European self-consciousness and a European purpose, provoked by the demonstration of the hitherto latent power of Asia; and, if a federation of European peoples is ever to be realised, it will be the result of their further development through opposition to a great and threatening Asiatic power, a revived Moslem empire, or possibly a threatened American domination.[1]

Although war has hitherto been the most important condition of the development of national consciousness, it is not the only one; and it remains to be seen whether industrial or other forms of rivalry can play a similar part. Probably, industrial rivalry cannot; the accumulation of wealth is too largely dependent upon the accidents of material conditions to become a legitimate source of national satisfaction; for, unlike the satisfaction arising from successful exertion of military power, it does not imply intrinsic superiorities. If the natural conditions of material prosperity could be equalised for all nations, then the acquisition of superior wealth, implying as it would superior capacities, might become a sufficiently satisfying end of national action; just as the equalisation of conditions among individuals in America has for the present rendered the accumulation of wealth a sufficient end, because such accumulation implies superior powers and is the mark of personal superiority.

Other forms of rivalry—rivalry in art, science, letters, in efficiency of social and political organisation, even in games and sports, all play some part; and it is possible that together they might suffice to constitute sufficient stimulus, even though the possibility of war should be for ever removed.[2]

But national self-consciousness is not developed by conflict and rivalry only. It is refined and enriched by all other forms of intercourse. In studying other peoples,

[1] This was written before the war with Germany.

[2] Emulation in the administration of backward peoples offers perhaps the greatest possibilities as "a moral equivalent for war."

their organisation and their history, we become more clearly aware of the defects and the qualities and potentialities of our own nation. And in this way, refinement of national consciousness is now going on rapidly in the European peoples. The latest considerable advance is due to the observation of Japan; for this has clearly demonstrated the imperfection of many conceptions that were current among us and has brought a certain abatement of national complacency and a greater earnestness of national self-criticism, which is highly favourable to increase of national self-knowledge.[1]

We might place nations in a scale of nationhood. The scale would correspond roughly to one in which they were arranged according to the degree to which the public good is the end, and the desire of it the motive, of men's actions; this in turn would correspond to a scale in which they were arranged according to the degree of development and diffusion of the national consciousness, of the idea of the nation or society as a whole; and this again to one in which they were arranged according to the degree of intercourse they have had with other nations. At the bottom of the scale would stand the people of Thibet, the most isolated people of the world; near them the Chinese, who also have until recently been almost entirely excluded from international intercourse. Such peoples have a national consciousness and sentiment which is extremely vague and imperfect. They do not realise their weakness, their strength, or their potentialities, but have an unenlightened pride without aspiration for a higher form of national life. A little above them would stand Russia, which has remained for so long outside the area of European international life. While at the top of the scale would be those nations which have borne their part in all the strain and stress and friction of European rivalry and intercourse.

[1] Cp. Principal L. P. Jacks on the Japanese in his *Alchemy of Thought.*

These degrees of international intercourse have been very largely determined by geographical conditions; isolation, and consequent backwardness in national evolution, being in nearly every case due to remoteness of position. The most important factor of modern times making for more rapid social evolution is probably the practical destruction or overcoming of the barriers between peoples; for thus all peoples are brought into the international arena, and their national spirit is developed through international intercourse and rivalry.

It is this increasing contact and intercourse of peoples, brought about by the increased facilities of communication, which has quickened the growth of national self-consciousness throughout all the world and has made the principle of nationality or, more properly, the desire for nationhood and for national existence and development, for self-assertion and for international recognition, the all-important feature of modern times, overshadowing every other phenomenon that historians have to notice, or statesmen to reckon with.

The American nation is interesting in this connection. If we ask—Why is their public life on a relatively low level, in spite of so many favouring conditions, including a healthy and strong public opinion?—the answer is that they have been until recently too much shut off from collective intercourse with other nations, too far removed from the region of conflict and rivalry. And judicious well-wishers of the American nation rejoice that it has recently entered more fully into the international arena, and has not continued to pursue the policy of isolation, which was long in favour; because, as is already manifest, this fuller intercourse and intenser rivalry with other nations must render fuller and more effective their national spirit, develop the national will and raise the national life to a higher plane, giving to individuals higher ends and motives than the mere accumulation of wealth, and removing that self-complacency as regards their national existence which hitherto has characterised them in common with the peoples of Thibet and China.

CHAPTER XII

Ideas in National Life

WE have seen that the idea of the nation can and does, in virtue of the formation of the sentiment of devotion to it, lead men to choose and decide and act for the sake of the nation; they desire the welfare and the good of the nation as a whole, they value its material prosperity and its reputation in the eyes of other nations; and, in so far as the decisions and actions of a nation proceed from this motive, co-operating with and controlling other motives in the minds of its members, such decision and action are the expressions of true collective volition.

It is truly volition, because it conforms to the true type of volition. Individual volition can only be marked off from merely impulsive action and every lower form of effort, by the fact that in true volition, among all the impulses or motives that may impel a man to action or decision, the dominant rôle is played by a motive that springs from his self-regarding sentiment. This motive is a desire to achieve a particular end, which, viewed as the achievement of the self, brings him satisfaction, because the thought of himself achieving this end is in harmony with the ideal of the self which he has gradually built up and has learnt to desire to realise under the influence of his social setting. The same is true of national volition.

And it is collective volition, in so far as the deliberations by which the decision of the nation has been reached have been effected through those formally and informally organised relations and channels of communication and by means of all the various modes of interaction of persons by which public opinion is formed and in which it is guided and controlled by the living traditions of the nation.

That this is the true nature of national volition may be more clearly realised on considering some instances of national action which could not properly be called the expression of the will of the Nation. A tariff might be adopted because a large number of men desired it, each in order that he himself might get rich more quickly; and, even though a large majority, or even all men, desired it, each for his private end, it would not be the expression of the national will, it would not be due to collective volition; it would be the expression of the will of all. Nor would it be an expression of the national will, even if each believed that, not only he, but also all his fellows would be enriched, and if he desired it for that reason also; that would be an expression of the will of all for the good of all. Only if and in so far as the decision was reached through the influence of those who desired it, because it seemed to them to be for the good of the whole nation, would it be the expression of the will of the nation.

And the difference would be not merely a difference of motive; the difference might be very important in respect both of the deliberative processes by which the decision was reached and also in respect of its ultimate consequences. For the will of all for the good of all would have reference only to the immediate future; whereas the truly national will would be influenced not only by consideration of the good of all existing citizens, but, in an even greater degree, by the thought of the continued welfare of the whole nation, in the remote future.

Again, suppose that, on the occasion of an insult or injury to the nation (I remind the reader of the incident in the North Sea when the Russian fleet fired on our fishing boats), a wave of anger against the offending nation sweeps over the whole country and that this outburst of popular fury plunges the nation into war. That would be collective mental process, but not volition; it would be action on the plane of impulse or desire, unregulated by reflection upon the end proposed in relation to the welfare of the nation and by the motives to action that are stirred by such reflection.

Again, suppose a nation of which every member was patriotic, and suppose that some proposed national action were pondered upon by each man apart in his own chamber, without consultation and discussion with his fellows in public and private. Then, though the decision would be true volition, in so far as it was determined by each man's desire for the national welfare, it yet would not be collective or national volition; because not reached by collective deliberation.

We have seen that the idea of the nation, present to the minds of the mass of its members, is an essential condition of the nation's existence in any true sense of the word "nation"; that the idea alone as an intellectual apprehension cannot exert any large influence; that it determines judgment and action only in virtue of the sentiment which grows up about this object—a sentiment which is transmitted and fostered from generation to generation, just because it renders the nation an object of value. The consideration should be obvious enough; but it has commonly been ignored by philosophers of the intellectualist school. They treat the individual mind as a system of ideas; they ignore the fact that it has a conative side which has its own organisation, partially distinct from, though not independent of, the intellectual side; and consequently they ignore equally the fact that the national mind has its conative organisation.

Imagine a people in whom anti-nationalism (in the form of cosmopolitanism, syndicalism, or philosophic anarchism) had spread, until this attitude towards the nation-state as such had become adopted by half its members, while the other half remained patriotic. Then there would be acute conflict and discussion, and the idea of the nation would be vividly present to all minds; but the nature of the sentiment attached to it would be different and opposite in the two halves; one of attachment and devotion in the one half; of dislike, aversion, or at least indifference (i.e. lack of sentiment) in the other half. And the efforts of the one half to maintain the nation as a unit would be antagonised and perhaps rendered nugatory by the indifference or opposition of the other half, who would always seek to break down national boundaries and would refuse co-operation in any national action, and who would league themselves with bodies of similar interests and anti-national tendencies in other countries. Then, even though all might be well-meaning people desiring the good of mankind, the nation would be very greatly weakened and probably would soon cease to exist as such.

The illustration shows the importance of the distinction which Rousseau did not draw in his discussion of the general will—namely, the distinction between the good of all and the good of the whole, i.e. of the nation as such. It might be argued that the distinction is purely verbal; it might be said that, if you secure the good of all, you thereby *ipso facto* secure the good of the whole, because the whole consists of the sum of existing individuals; and that this is obvious, because, if you take them away, no whole remains. But to argue thus is to ignore the fact on which we have already insisted—namely, that the whole is much more than the sum of the existing units, because it has an indefinitely long future before it and a part to play, through indefinitely long periods of time, as a factor in the general welfare and progress of mankind.

So much greater is the whole than the sum of its existing parts, that it might well seem right to sacrifice the welfare and happiness of one or two or more generations, and even the lives of the majority of the citizens, if that were necessary to the preservation and future welfare of the whole nation as such. This is no merely theoretical distinction; it is one of the highest practical importance, which we may illustrate in two ways.

A whole nation may be confronted with the alternative, may be forced to choose between the good of all and the good of the whole. Such a choice was, it may be said without exaggeration, suddenly presented to the Belgian people, and only less acutely to ourselves and to Italy, by the recent European conflagration; and in each case the good of the whole has been preferred. Is it not probable and obvious that, if each or all of these peoples had consented to the domination of Germany, the material welfare of all their existing citizens might well have been increased, rather than diminished, and that their choice has involved not only the loss of life of large numbers of their citizens and great sufferings for nearly all the others, but also enormous sacrifice of material prosperity, in order that the whole may survive and eventually prosper as a nation working out its national destiny free from external domination? There are, or were, those who say that they would just as soon live under German rule, because they would be governed at least as well and perhaps better than by their own government hitherto; and there is perhaps nothing intrinsically bad or wrong in this attitude; the question of its rightness or wrongness turns wholly on the valuation of nationality. It is easier to appreciate this plea on behalf of another people than our own. One may hear it said even now that, after all, it would have been better for Belgium that she should have entered into the group of Germanic powers in some sort of federal system or Customs union; that, in general, it is ridiculous that the small states should claim sovereign powers and pretend to have their own foreign policy and so forth; that they are struggling against the inevitable, against a universal and necessary tendency for the absorption of the smaller states by the larger.

We may illustrate the difference between regard for the good of all members of the nation and of the nation itself in another way—namely, by reference to socialism, the principle which would abolish inequalities of wealth and opportunity, as far as possible, by abolishing or greatly restricting the rights of private property and capital, especially the right of inheritance. There can, I think, be little doubt that the adoption of socialism in this sense by almost any modern nation would increase the well-being and happiness of its members very decidedly on the whole for the present generation and possibly for some generations to come. It is in respect of the continued welfare of the whole and of its perpetuation as an evolving and progressing organism that the effects seem likely to be decidedly bad. The socialists are in the main those who fix their desire and attention on the good of all; hence they are for the most part inclined to set a low value on nationality, even while they demand a vast extension of the functions of the State, conceived as an organised system of administration. Those, on the other hand, who repudiate socialism, not merely because they belong to the class of "Haves," must seek their justification in the consideration of the probable effects of such a change on the welfare of the nation conceived as an organism whose value far transcends the lives of the present generation.

When, then, we attribute to the idea of the nation or to the national consciousness this all-important creative, constitutive, and conservative function, we must be clear that the idea is not an intellectual conception merely, but implies an enduring emotional conative attitude which is the sentiment of devotion to the nation; and, further, we must remember that the nation means not simply all existing individuals, the mere momentary embodiment of the nation, but something that is far greater, because it includes all the potentialities embodied in the existing persons and organisation.

It is the presence and operation in the national mind of the idea of the nation in the extended sense just indicated that gives to national decisions and actions the character of truly collective volitions; they approach this type more nearly, the more the idea is rich in meaning and adequate or true, and the more widely it is spread, and the more powerful and widely spread is the sentiment which attaches value to the nation and sways men to decision and action for the sake of the whole, determining the issue among all other conflicting motives.

And it is the working of the national spirit and the acceptance of and devotion to the national organisation which render the submission of the minority to the means chosen by the majority a voluntary submission; for it is of the essence of that organisation that, while all accept and will the same most general end, namely the welfare of the whole, the choice of means must be determined by the judgment of the majority, formed and expressed as a collective judgment and opinion by way of all the many channels of reciprocal influence that the national organisation, both formal and informal, provides. In so far as each man holds this attitude, esteeming the nation and accepting loyally its constitution or organisation, the decision determined by even a bare majority vote of parliament becomes the expression of the national will; and the co-operation in carrying it out of those who did not judge the method to be wise, and who therefore voted against it, yet becomes a truly voluntary co-operation, in so far as they accept the established organisation.

The point may be illustrated by the instance of a nation going to war. A large minority may be against war, for reasons which to them may seem to be of the highest kind; it may be that they judge the nation to be morally in the wrong in the matter in dispute, or very questionably in the right, as many Englishmen did during the Boer War; and yet, if, by the accepted organised channels of national deliberation and decision, war has been declared, then, although it was their duty to do what they could to make their opinion prevail before the decision was reached, there is no moral inconsistency in their supporting the war measures with all their strength. It is in fact implied in their loyalty, if they are loyal and patriotic, that they shall yield their individual opinion to the expression of the national will and shall accept the means chosen to the common end. That is the truth implied in the phrase—My country right or wrong. Of course, this phrase may be taken in a reprehensible sense, as meaning that any opportunity of forwarding the immediate interests of one's country must be taken, regardless of the interests of other communities and of the obligations of common honesty and humanity upon which all human welfare depends.

In the same way, a man might disapprove of a particular tax, say on liquor, or of obligatory military service; and yet he may accept the national will and serve faithfully as a soldier, without inconsistency, and without ceasing to be a free agent truly willing the acts imposed by his position in the whole organisation; just as during the late war many priests served as soldiers in the French army. Or, to take an extreme instance, a man who has broken the law and even incurred the death penalty may be truly said to undergo his punishment of imprisonment or death as a morally free agent, if he is loyal to his country and its institutions, accepting the penalty, while yet believing his action to be right. Such perfect loyalty to the nation is of course rare; and in all actual nations men have progressed towards it in very different degrees. Most existing nations have emerged from preceding despotisms by the repeated widening of the sphere of freedom, as the growth of loyalty in strength and extension rendered such freedom consistent with the survival of the State and its administrative functions.

Thus a people progresses from the status of an organism, in which the parts are subordinated to the whole without choice or free volition on their part, or even against their wills, towards the ideal of a Nation-State, an organic whole which is founded wholly upon voluntary contract between each member and the whole, and in which the distinction between the State and the nation becomes gradually overcome and replaced by identity. For, as national self-consciousness develops and each man conceives more fully and clearly the whole nation and his place and function in it, and grows in loyalty to the nation, he ceases to obey the laws merely because he is constrained by the authority and force of the State. An increasing proportion of citizens obey the law and render due services voluntarily, because they perceive that, in so doing, they are contributing towards the good of the whole which they value highly; in so far as they act in this spirit, the actions and restraints prescribed by law become their voluntary actions and restraints.

Thus the theory that society is founded upon a *Social Contract*, which, if taken as a description of the historical process of genesis, is false, is true, if accepted as the constitutive principle of the ideal State towards which progressive nations are tending.

And, as the organisation of a nation becomes less dependent upon outer authority and upon mere custom and the unreasoning acceptance of tradition, and more and more upon free consent and voluntary contract, the nation does not cease to be an organism; it retains that formal and informal organisation which has developed in large part without the deliberate guidance of the collective will and which is essential to its collective life; the national mind, as it grows in force and extension and understanding of its own organisation, accepts those features which it finds good, and gradually modifies those which appear less good in the light of its increasing self-knowledge; and so it tends more and more to become a contractual organism, which, as Fouillée has insisted, is the highest type of society.

It should be noticed that this ideal of the contractual organism synthesises the two great doctrines or theories of society which have generally been regarded as irreconcilable alternatives: the doctrine of society as an organism, and that of society as founded upon reason and free will. They have been treated as opposed and irreconcilable doctrines, because those who regarded society as an organism, taking the standpoint of natural science, have laid stress upon its evolution by biological accidents and by the interaction and conflict of many blind impulses and purely individual volitions, in which collective volition, governed by an ideal of the form to be achieved, had no part. While, on the other hand, the idealist philosophers, describing society or the nation as wholly the work of reason and free will, have been guilty of the intellectualist fallacy of regarding man as a purely rational being; they have ignored the fact that all men, even the most intellectual, are largely swayed and moulded by processes of suggestion, imitation, sympathy, and instinctive impulse, in quite non-rational ways; and they have ignored still more completely the fact that the operation of these non-rational processes continues to be not only of immense influence but also inevitable and necessary to the maintenance of that organic unity of society upon which as a basis the contract-unity is superimposed as a bond of a higher, more rational, and more spiritual quality.

The former doctrine logically tends to the paralysis of social effort and to the adoption of extreme individualism, to the doctrine of each man for himself, and of *laissez faire*, doctrines such as those of Herbert Spencer. The other, the idealist theory of the State as being founded and formed by reason, tends equally logically towards extreme State socialism; because its overweening belief in reason leads it to ignore the large and necessary basis of subrational organisation and operation.

Only a synthesis of the two in the doctrine of the contractual organism can reconcile them and give us the ideal of a nation in which the maximum and perfection of organisation shall be combined with the maximum of liberty; because in it each individual will be aware of the whole and his place and functions in it, and will voluntarily accept that place and perform those functions.

The highest, most perfectly organised and effective nation is, then, not that in which the individuals are disposed of, their actions completely controlled, and their wills suppressed, by the power of the State. It is, rather, one in which the self-consciousness and initiative and volition of individuals, personality in short, is developed to the highest degree, and in which the minds and wills of the members work harmoniously together under the guidance and pressure of the idea of the nation, rendered in the highest degree explicit and full and accurate.

The Value of Nationality

At the present time, while the mass of men continue to accept the duty of patriotism unquestioningly, and historians for the most part are content to describe with some astonishment the immense development of nationalism in the past century, many voices are loudly raised for and against nationality. The great mass of men no doubt are swept away in the flood of patriotic feeling. But the war has also intensified the antipathy, and given increased force to the arguments, of those who decry nationality and deprecate patriotism—for these are but two different modes of expressing the same attitude.

There are two principal classes of the anti-nationalists. First, the philosophic anarchists, who would abolish all states and governments, as unnecessary evils, men like Kropotkin and Tolstoi. Secondly, the cosmopolitans, who, while believing in the necessity of government and even demanding more centralised administration, would yet abolish all national boundaries as far as possible, boundaries of geography, of language, race, and sentiment, and all national governments, and would aim at the establishment of one great world-state.

Though the aims of these two parties are so widely different, they use much the same arguments against nationalism. According to the anti-nationalist view, nationalism and the patriotism in which it is founded are a kind of disease of human nature, which, owing to the unfortunate fact that mankind has retained the gregarious instinct of his animal ancestors, inevitably breaks out as soon as any community begins to come into free contact and rivalry with other communities, and which tends to grow in force in a purely instinctive and irrational manner the more these contacts and rivalries increase.

The liability to patriotism is thus regarded as closely comparable with mankind's unfortunate liability to drunkenness, to feel the fascination of strong liquor—as merely a natural and inevitable result or by-product of an unfortunate flaw in human nature—a tendency which will have to be sternly repressed and, if possible, eradicated, before men can hope to live in peace and tolerable security and to develop their higher capacities.

The fact that patriotism of some degree and form is universally displayed, and that it breaks out everywhere into heat and flame when certain conditions are realised, does not for them in any degree justify it; and it should not, they hold, reconcile us to its continued existence; they draw an indictment not merely against a whole people, but against the whole human race. They attack nationalism, firstly, by describing what, in their opinion, patriotism is and whence it comes; secondly, by describing what they believe to be the natural consequences and effects of nationalism.

The most common mode of attack is to identify patriotism with jingoism; they speak of "jelly-bellied flag-flappers," of flag-wagging and mafficking; they assert that the essence of patriotism is hatred and all uncharitableness towards other countries and their citizens.[1]

Less virulent is the criticism of those who, looking coldly upon patriotism, describe it as the mere blind expression of the working of the gregarious instinct among us, and as something therefore quite irrational, which must and will tend to disappear, as men become more enlightened and are guided more by reason and less by instinct.[2]

Again, it is said that patriotism is a form of selfishness and therefore bad; that it is a limitation of our sympathies, a principle of injustice; that it stands in the way of the realisation of universal justice, of the universal brotherhood of man, which is the ideal we obviously must accept

[1] W. L. George, *English Review*, May, 1915, "The Price of Nationality." "Anger, indeed, is the soul of what is called the national will. To call it a will is perhaps too much, it is an instinct and mainly an instinct to hate. . . . Love of country is mainly hatred of other countries."

[2] Cf. Gilbert Murray, *Collection of Addresses on The War given at Bedford College*, 1915.

and aim at. Or in other words, and this is the main indictment, it is alleged that patriotism and nationalism inevitably tend to produce war, that they keep the rival nations perpetually arming for possible wars and actually in commercial and economic war, if not at real war. And of course the evils of warfare and of such perpetual preparation for war are great and obvious enough in modern Europe. In support of this indictment, they point to the golden age of the Roman Empire, when the inhabitants of all its parts were content to sink their differences of race and country and were proud to proclaim themselves citizens of the Roman Empire; and they say that in consequence the civilised world attained then a pitch of prosperity and contentment never known before or since over any large area of the earth.

This is a formidable indictment, to which the exponents and advocates of patriotism have for the most part been content to reply by renewed exhortations to patriotism, by emotional appeals, by rhetoric, by the quotation of patriotic verses, the citation of the glorious deeds of our armies and soldiers now and in past times, by all the arts of persuasion and suggestion. As a fine example of this method one may cite Mr. Stratford Wingfield's *History of British Patriotism*, in which he not only confines himself to these methods, but shows a positive dislike and contempt for all attempts to apply reason and scientific method to the study of human affairs.[1] In maintaining this attitude, the advocates of patriotism give some colour to the claim

[1] Incidentally he holds up my *Social Psychology* as a dreadful example of such an attempt and a woeful evidence of the parlous state of present-day culture in England. Such dislike of any attempt to understand that which we hold sacred is intelligible enough in the vulgar, for whom all analysis is destructive of the values they unreasoningly cherish. But it may be hoped that men of letters who set out to defend patriotism will learn to rise above this attitude, just as the more enlightened leaders of religion are learning to welcome psychological inquiry in their domain.

of their opponents that patriotism or nationalism is essentially irrational, in the sense that it is incapable of justification by reason.

The politicians and historians, on the other hand, who are so generally demanding that the European settlement after the war must accept nationality as its fundamental principle, are commonly content to note the strength and the wide distribution of the patriotic sentiment, without enquiring into its origin, nature, or value.

Let us examine the arguments against patriotism, and then see what reason can advance in its defence. For, though a rational defence of patriotism will have little direct effect in making patriots, we may be sure that, if such defence cannot be maintained, patriotism will have to fight a losing battle.

In disparaging patriotism by describing it as the work of an instinct, the gregarious or the pugnacious or other instinct, or of several instincts, its critics are guilty of two psychological errors and a popular fallacy. The last is the fallacy that the worth of any thing is to be judged by the source from which it springs. Even if patriotism were nothing more than the direct expression of the gregarious instinct, which we possess in common with many of the higher animals, that would not in itself condemn it. But this description of it, as a product of instinct as opposed to the principles we attain by reason, involves that false disjunction and opposition of reason to instinct which is traditional and which the intellectualist philosophers commonly adopt, when they condescend to recognise in any way the presence of instinctive tendencies in human nature.

The other psychological error is the failure to recognise that patriotism, although, like all other great mental forces, it is rooted in instinct, is not itself an instinct or the direct expression of any instinct or group of instincts, but is rather an extremely complicated sentiment, which has a long and complex history in each individual mind in which it manifests itself; that it is, therefore, capable of infinite variety and of an indefinite degree of intellectualisation and refinement; that the cult of patriotism is, therefore, a field for educational effort of the highest order, and that in this field moral and intellectual education may achieve their noblest and most far-reaching effects.

The psychological justification of patriotism has already been indicated, but may be concisely stated here. The moral value of the group spirit was considered in an earlier chapter; we saw how it, and it alone, raises the conduct of the mass of men above the plane of simple egoism or family selfishness. The sentiment of devotion or loyalty to any group has this virtue in some degree; but loyalty to the nation is capable of exalting character and conduct in a higher degree than any other form of the group spirit. For the nation alone has continuity of existence in the highest degree; a long past which gives a large perspective of past history, involving the history of long series of self-sacrificing efforts and many heroic actions; and the prospect of an indefinitely prolonged future, with the possibility of continued progress and development of every kind, and therefore some security for the perpetuation of the results achieved by individual efforts.[1]

Further, the nation alone is a self-contained and complete organism; other groups within it do but minister to the life of the whole; their value is relative to that of the whole; the continuance of results achieved on their behalf is dependent upon the continued welfare of the whole (for example, the welfare of any class or profession—a fact too easily overlooked by those in whom class spirit grows strong). Hence, the nation, as an object of sentiment,

[1] In these respects the Church alone can enter into serious rivalry as an object of loyalty.

includes all smaller groups within it; and, when the nation is regarded from an enlightened point of view, the sentiment for it naturally comes to include in one great system all minor group sentiments, and to be strengthened by their incorporation.

It is important to notice also that, just as the minor group sentiments are not incompatible with, but rather may strengthen, the national sentiment, when subordinated to and incorporated in it, so the national sentiment is not incompatible with still more widely inclusive group sentiments—for example, that for a European system of nations, for the "League of Nations" or for Western Civilisation in general. And, while loyalty to humanity as a whole is a noble ideal, it is one which can only be realised through a further step of that process of extension of the object of the group sentiment, of which extension patriotism itself is the culmination at present for the great mass of civilised mankind. The attempt to achieve it by any other road is bound to fail, because psychologically unsound.[1]

Let us note in passing that neglect of this truth gives rise to two of the extreme forms of political doctrine or ideal, current at the present day; first, the ideal of the brotherhood of man in a nationless world; secondly, the extreme form of democratic individualism which assumes that the good of society is best promoted by the freest possible pursuit by individuals of their private ends, which believes that each man must have an equal voice in the government of his country, because that is the only way in which his interests and those of his class can be protected and forwarded; a doctrine which regards public life as a mere strife of private and class interests. Both ideals fly in the face of psychological facts; and, though

[1] As Dean Inge has remarked—"If they love not those whom they have seen, how shall they love those whom they have not seen?"

they are in appearance extreme opposites, they are apt to be found associated in the same minds.

At the other end of the scale, we have the philosophical conservatism of such a thinker as Edmund Burke, which is keenly aware of the organic unity of society and looks constantly to the good of the whole, deriving from that consideration its leading motives and principles, and which trusts principally to the growth of the group spirit for the holding of the balance between conflicting interests and for the promotion of the public welfare.

Having seen the importance for national life of the idea of the nation, the diffusion of which through the minds of the people constitutes national self-consciousness, let us glance for a moment at the way other ideas may play leading roles in national life. Such are ideas which became national ideals, that is to say, ideas of some end to be realised by the nation which become widely entertained and the objects of strong sentiments and of collective emotion and desire and which, therefore, determine collective action.

I shall not attempt to deal separately with various classes of such ideas, or ideals—the political, the religious, the economic; but shall only note the fact that they have played, and may yet play, great parts in the history of the world.

Men are not swayed exclusively by considerations of material self-interest, as the older school of economists generally assumed; nor even by spiritual self-interest, as too much of the religious teaching of the past has assumed; nor even by consideration of the welfare of the social groups of which they are members. Many of the great events of history have been determined by ideas that have had no relation to individual welfare, but have inspired a collective enthusiasm for collective action, for national effort, of a distinterested kind; and the lives of some nations have been dominated by some one or two such ideas. These ideas are first conceived and taught by some great man, or by a few men who have acquired prestige and influence; they then become generally accepted by suggestion and imitation, accepted more or less uncritically and established beyond the reach of argument and reasoning.

No matter what the character of the idea, its collective acceptance by a people enhances for the time the homogeneity of mind among them, renders the people more intimately a unity, and serves also to mark it off more sharply from other peoples among whom other ideas prevail.

But, besides thus binding together at any period of its history the people that entertains it, the generally accepted idea, if it endures, may produce further effects by becoming incorporated in the national organisation; in so far as it determines the form of activity of the people, it moulds their institutions and customs into harmony with itself, until they become in some measure its embodiment and expression; and in any vigorous nation there are usually one or two dominant ideas at work in this way.

It is a favourite dogma with some writers (for example M. le Bon) that ideas, before they can exert great effects in the life of a nation, must first become unconscious ideas, incorporated as they say in the unconscious soul of a people. This is an obscure confused doctrine, which, if it is meant to be taken literally, we can only reject. If it is to have any real meaning, it must be taken in the sense that the long prevalence of the ideal moulds the institutions and customs and the executive organisation of a people, so that national action towards the ideal end becomes more or less automatic or routine.

If the ideal so accepted and incorporated in the organised structure of the national mind, is one that makes for strength and at the same time permits of progress, it lives on; in other cases it may destroy the nation, or petrify it, arresting all progress.

Consider one or two examples of ideas that have played dominant roles in the lives of nations. They are mostly political, or religious, or, most powerful of all, politico-religious. The idea of world-conquest has dominated and has destroyed several great nations, of which the latest example is the German Empire. The idea of conversion by the sword, accepted with enthusiasm by the Arab nation, gave it for a time tremendous energy, but contained no potency of permanent power or of progress. The idea of immortality, or desire of continued existence after death, seems to have dominated the minds of the ancient Egyptian people; the idea of escape from the evils of this world, those who have fully accepted Buddhism, like the Burmans.[1] The idea of caste as an eternal and impassable barrier has largely determined the history of India.

All these are ideas which have proved ineffective to sustain national vigour or to promote social evolution. It would not be strictly true to say that the fall, or the unprogressive condition, of the peoples that have entertained these ideas is the result of those ideas; because the general acceptance of them proves that they were in harmony with the type of mind of the people. Yet the formulation of the ideas by the leading minds who impressed them on the peoples must have accentuated those tendencies with which they harmonised; and in each case, if the idea had never been formulated, or if others had been effectively impressed on the mind of the people, the course of its history would have been changed.

Of ideas less adverse to national life take the idea of ancestor worship, and the idea of personal loyalty to the ruler, ideas which commonly go together and have played

[1] Cp. Fielding Hall, *The Soul of a People*.

an immense part in the life of some peoples, notably in Japan; they have served as effective national bonds in periods of transition through which despotically ruled populations have progressed to true nationhood. The idea of the divine right of kings played for a time a similar role in Europe.

A good example of the operation of an ideal in a modern nation is that of the ideal of a great colonial empire in the French nation. No doubt, hopes of economic advantages may have played some part in this case; but the growth of the immense oversea empire of modern France, as well as of the great extra-European conquests which France has made in the past but has ceased to control, seems to have been due in the main to the operation of this ideal in the national mind. France has no surplus population, and no Frenchman desires to leave his beautiful France; everyone regards himself as cruelly exiled if compelled to live for a time in any of the oversea possessions; and most of these, notably the Indo-Chinese Empire, are very expensive, costing the nation far more in administrative expenses than any profits derived from them, and involving constant risks of international complications and war, as in Morocco in recent years. Nevertheless, the ideal still holds sway and, under its driving power, the oversea territories of France, especially in Africa, have grown enormously. And this ideal has inevitably incorporated itself in the organisation of the nation, in a colonial office and a foreign legion, and all the administrative machinery necessarily set up for securing the ends prescribed by the ideal.

In modern times the most striking illustration of the power of ideas on national life is afforded by the influence of the ideals of liberty and equality. It was the effective teaching of these ideals of liberty and equality, primarily by Rousseau, to a people prepared by circumstances to receive them, which produced the French Revolution; and all through the nineteenth century they have continued to determine great changes of political and social organisation in many countries of Europe and in America.

In England the idea of liberty has long been current and long ago had become incorporated and expressed in the national organisation; but its application received a vast extension when in 1834 England insisted on the liberation of all British-owned slaves and paid twenty million sterling in compensation. That the idea still lives on among us, with this extended application, seems to have been proved by the results of recent elections which were influenced largely by the force of the no-slavery cry in relation to coloured labour. It is an excellent example of an established collective ideal against which reason is of no avail.

The ideal of liberty never entered the minds of the most advanced peoples of antiquity; their most enlightened political thinkers could not imagine a State which was not founded upon slavery. Yet it has become collectively accepted by all the leading nations; and the ordinary man has so entirely accepted it that he cannot be brought to reason about it. Facts and arguments tending to show that the greater part of the population of the world might be happier without liberty and under some form of slavery cannot touch or enter his mind at all.

The ideal of political equality is of still later growth, and is in a sense derivative from that of liberty; it was in the main accepted as a means to liberty, but has become an end in itself. It is moulding national organisation everywhere; through its influence parliamentary government and universal suffrage are becoming the almost universal rule; and, through leading to their adoption, this ideal is in a fair way to wreck certain of the less firmly organised nations, and possibly our own also.

But the ideal which, beyond all others, characterises the

present age of almost all the nations of the world is the ideal of progress. Hardly anyone has any clear notion of what he means by progress, or could explicate the idea; but the sentiment is very strong, though the idea is very vague. This idea also was unknown to the leading thinkers of antiquity and is of recent growth; yet it is so almost universally accepted, and it so permeates the mental atmosphere in every direction, that it is hard for us to realise how new a thing in the history of the world is the existence, and still more the effective dominance, of the idea. It is perhaps in America that its rule is most absolute; there the severest condemnation that can be passed by the average man upon any people or institution is to say that it is fifty years behind the time. The popular enthusiasm for flying-machines, which threatens to make life almost unlivable, is one of the striking illustrations of the force of this ideal.

More recent still, and perhaps equally important, is the idea of the solidarity of the human race and of the responsibility of each nation towards the rest, especially towards the weaker and more backward peoples. We no longer cheerfully and openly exterminate an inferior people; and, when we do so, it is with some expressions of regret and even of indignation.

But this moral idea is still in process of finding acceptance and illustrates well that process. It has been taught by a few superior minds and none dares openly repudiate it; hence, it gains ground and is now commonly accepted, verbally at least, and is just beginning to affect national action.

The four ideas, liberty, equality, progress, and human solidarity or universal responsibility, seem to be the leading ideas of the present era, the ideas which, in conjunction with national sentiments, are more than any other, fashioning the future of the world.

The last two illustrate exceptionally well the capacity of nations to be moved by abstract ideas not directly related to the welfare of the individuals whose actions they determine; they show once more how false is the doctrine that national life is but the conflict of individual wills striving after individual good. They show that, through his life in and mental interaction with organised society, man is raised morally and intellectually high above the level he could individually achieve.

CHAPTER XIII

Nations of the Higher Type

LET us consider now the type of nation which from our present point of view is the most interesting, the type which approximates most nearly to a solution of the problem of civilisation, to the reconciliation of individuality with collectivity, to the synthesis of individualist and collectivist ideals; that in which the rights and wills of individuals are not forcibly subordinated to those of the State by the power of a governing class, and in which the deliberative side of the national mind is well developed and effective.

Such are in a certain degree the French, but still more the British and the American nations. In the two latter countries the rights of the individual are made supreme over all other considerations, the welfare of the whole is only to be advanced by measures which do not override individual wills and rights; or, at least, the only power which is admitted to have the right in any degree to override individual wills is the will of the majority. In such a nation the greatest efforts are concentrated on the perfection of the deliberative organisation, by means of which the general mind may arrive at collective judgment and choice of means and may express its will. A vast amount of time and energy is devoted to this deliberative work; while the executive organisation, by which its decisions have to be carried into effect, is apt to be comparatively neglected and hence imperfect.

These two complementary features of such States we see well exemplified here and in America;[1] where the amount of time, money, and effort spent upon the deliberative processes and the elaboration of the organisation through which they are effected is enormously greater than in other nations. And, in spite of the energy expended on deliberative processes and on the elaboration of their organisation, the interests of the nation as a whole are not at present forwarded in a manner at all comparable with those of such a State as Germany. Nevertheless, such national actions as we do achieve are far more truly the expression of the national will; and, if the national mind is to be developed to a high level, this vast expenditure of energy, which to some impatient spirits seems wasteful and useless, must go on.

As was said in a former chapter, such collective deliberation of modern nations is only rendered possible by the great facilities of communication we enjoy; telegraph, post, and railway, and especially the press. The ancients saw, truly enough that, with their limited means of communication, the higher form of State-organisation must be restricted to a small population of some thousands only—the City-State.

It is important to note that not only do modern facilities of communication render possible a truly collective mental life for the large Nation-States of the present age; but that these modern conditions actually carry with them certain great advantages, which tend to raise the collective mental life of modern nations to a higher level than was possible for the ancient City-State, even though its members were of high average capacity and many of them of very great mental power, as in Athens.

[1] This, as President Lowell clearly shows in his *Public Opinion and Popular Government*, is carried to an extreme in America, and lies at the root of many administrative evils.

The assembly of citizens in one place for national deliberation rendered them much more susceptible to those less desirable peculiarities of collective mental life which characterise simple crowds; particularly, the excess of emotional excitement, increased suggestibility, and, hence, the ease with which the whole mass could be swayed unduly by the skilful orator. In the modern nation, on the other hand, the transmission of news by the press secures a certain delay, and a lack of synchronism, in its reception by different groups and individuals; and it secures also a certain delay in the action and reaction of mind on mind, which gives opportunity for individual deliberation. Also the sympathetic action of the mass mind on the individual mind is in large part indirect, rather than direct, representative rather than perceptual, and therefore less overwhelming in its effects. These conditions greatly temper the violence of the emotional reactions and permit of a diversity of feeling and opinion; an opposed minority has time to form itself and to express an opinion, and so may temper the hasty and emotional reaction of the majority in a way that is impossible in a general assembly.

A further advantage of the large size of nations may arise from the fact that actual decision as to choice of means for effecting national action has to be achieved by means of representatives who come together in one place. Representative government is not merely an inferior substitute for government by general assembly; it is superior in many respects. If each representative were a mere delegate, an average specimen of the group he represents, chosen by lot and merely charged to express their will, this feature would modify the crude collective mental processes in one important respect only; namely, it would counteract to some extent that weakening of individual responsibility which is characteristic of collective mental action. But, in addition to this, internal organisation, in the form of tradition and custom, comes in to modify very greatly the collective process.

We see such modifying influence very clearly in the election of the English House of Commons and in the methods of its operations. Owing partly to a natural tendency, partly to a fortunate tradition, the people do not elect just any one of themselves to serve as a delegate or average sample of the mass; but as a rule they choose, or try to choose, some man who displays special capacity and special qualifications for taking part in the national deliberations. In so far as they are successful in this, their representatives are able men and men to whose minds the social consciousness, the consciousness of the whole people, of its needs and tendencies and aspirations, is more fully and clearly present than to the average mind. They are also in the main men of more than average public spirit. Hence it is not unknown that a purely working class constituency, being offered liberal, conservative, and labour candidates, instead of choosing the labour man, one of themselves, gives him only a small fraction of the total votes. Then, within the body of representatives, this process, by which greater influence is given to the able men, to those whose minds reflect most fully the whole people, is carried further still. A small group of these men exerts a predominant influence in all deliberations; and not only are they in the main the best qualified (for they only attain their leading positions by success in an intense and long continued competition) but they are put in a position in which they can hardly fail to feel a great responsibility resting upon them; and in which they feel the full force of political traditions. The deliberative organisation of the American nation illustrates, when compared with our own, the importance of these traditions; for its lesser efficiency is largely due to the absence of such traditions, and to the fact that their system banishes from the House of Representatives its natural leaders and those on whom responsibility falls most heavily.

Lastly, the existence of two traditionally opposed parties ensures that every important step shall be fully discussed. The traditional division into two parties, which from one point of view seems so irrational, nevertheless exerts very important and valuable influences, of which the chief is that it prevents the assembly of legislators becoming a mere psychological crowd easily swayed to a decision by collective emotion and skilful suggestion; for each suggestion coming from the one party acts by contra-suggestion upon the other and provokes an opposition that necessitates discussion.[1]

In these two ways, then;—first, through the culmination of national deliberation among a selected group of representatives, among whom again custom and tradition accord precedence and prestige to the natural leaders, the most able and those in whose consciousness the nation, in the past, present, and future, is most adequately reflected; secondly, by means of the party system, which ensures vigorous criticism and full discussion of all proposals, under a system of traditional conventions evolved for the regulation of such discussions—in these two ways the principal vices of collective deliberation are corrected, and the formal deliberations and decisions of the nation are raised to a higher plane than the collective deliberations of any assembly of men lacking such traditional organisation could possibly attain. The part played by unwritten tradition in the working of the British constitution is of course immense, as for example, the existence

[1] President Lowell (*op. cit.*) has clearly shown other benefits resulting from the party system; he shows especially how the party is needed to prepare a program and select candidates, if the popular vote is to give expression to the dominant opinion of the people.

and enormous prestige of the cabinet, and the tradition that a party coming into power must respect the legislation of the party previously in power. Without this last, representative government, or at any rate the party system, would be impossible. The smooth working of the system depends entirely upon the influence of these and similar traditions which exist only in the minds of men. Or, take as another example, the tradition of absolute impartiality on the part of the Speaker and of loyal acceptance of his rulings by every member of the House; or the tradition which distinguishes sharply between political and private relations, in virtue of which the parties to a most bitter political strife may and very generally do remain in perfectly friendly private relations.

These and other such traditions, which secure the efficient working of the organisation for national deliberation, all rest in turn upon a traditional and tacit assumption—namely, the assumption that both parties are working for the good of the nation as they conceive and understand it, that both parties have this common end and differ only in their judgment as to the means by which it can best be achieved. They rest also on the traditional and tacit admission that one's own judgment, and that of one's party, may be mistaken, and that in the long run the legislation which any party can effect is an expression of the organised national mind and is therefore to be respected. It is this acquiescence in accomplished legislation in virtue of this tacit assumption which gives to the decisions of Parliament the status, not merely of the expression of the will of a bare majority, but of the expression of the will of practically the whole nation. Underlying the stability of the whole system, again, is the tradition, sedulously fostered and observed by the best and leading minds, that the *raison d'être* and purpose of the representative parliament is to organise, and to give the most complete possible expression to, the national mind and will; and that no constitutional change or change of procedure is justifiable unless it tends to the more complete realisation of these objects.

In virtue of these traditions our Parliament and Press constitute undoubtedly the best means for effecting organisation of the national mind in its deliberative aspect that has yet been evolved; and we should remember this when we feel inclined to gird at the "great talking shop" at the slowness of its procedure and at the logical absurdities of the two-party system; and, above all, we should realise how valuable and worthy of conservation are these scarcely formulated traditions, for they are absolutely essential to its efficiency. It is just because the efficiency of the deliberative organisation of a nation depends upon the force of such traditions, that, though it is possible to take the system of parliamentary representation and establish it by decree or plebiscite in a nation which has hitherto had no such deliberative organisation, it is not possible to make it work smoothly and efficiently amongst such a people. Hence, although almost every civilised nation has done its best to imitate the British system of parliamentary government, hardly any one has made a success of it; and, in nearly all, it is in constant danger of being superseded by some more primitive form of government—one need only mention Mexico, Portugal, Russia, France, Austria-Hungary. In all these countries, and even in America, there seems to be already a not very remote possibility of the supersession of parliamentary government by a dictatorship—a process which has actually occurred in many of the municipal governments of America, and the fear of which has constantly checked the smooth working of the parliamentary system in France.

As a single illustration of the way in which the conditions we have been considering affect the collective acts of the nation, consider what happened at the time of the Russian outrage in the North Sea during the Russo-Japanese war. When a Russian fleet fired upon our fishing boats doing considerable damage to them, the means of communication were sufficiently developed among us to allow of the action and reaction of all on each which produces the characteristic results of collective mental action, the exaltation of emotion, the suggestibility, the sense of irresponsible power; and, in the absence of the deliberative organisation which, by concentrating influence and responsibility in the hands of a few of the best men, controlled and modified this collective action, we should have rushed upon the Russian fleet and probably have brought on a general European war. The control and counteraction of this kind of outburst of collective emotion and impulsive action is one of the heaviest responsibilities of those to whom predominant influence is accorded.

It is only in virtue of the strong organisation of the national mind resting upon these long traditions of parliamentary government, that at such a time control of the popular emotion and impulse is possible. And the weaker and less efficient is such traditional organisation, the more does any such incident tend to provoke a collective manifestation which approximates in its uncontrollable violence and unconsidered impulsiveness to the behaviour of an unorganised crowd. Hence governments, where the democratic principle is acknowledged but the traditional organisation is less strong, are constantly in danger of having their hands forced by some outburst of popular passion—as in France.

It is worth noting that, when Aristotle inveighed against democracy as an evil form of government, the only form of democratic government he had in mind was government by the voices of a mob gathered together in one place and lacking all the safeguards which, as we have seen, render our British national deliberations so much superior to those of a mere crowd of persons of equally good average capacity and character.

But it is not only in the formal deliberations of the nation that internal organisation, resting on tradition, secures the predominance of the influence of the best and ablest minds. The same is true of all national thought and feeling. There exists in every great nation the vague influence we call public opinion, which is the great upholder of right and justice, which rewards virtue and condemns vice and selfishness. Public opinion exists only in the minds of individuals (for we have rejected, provisionally at least, the conception of a collective consciousness) yet it is a product not of individual, but of collective, mental life. And it has in any healthy nation far higher standards of right and justice and tolerance than the majority of individuals could form or maintain; that is to say, it is in these respects far superior to an opinion which would be the mere resultant or algebraic sum of the opinions of all the living individuals. In reference to any particular matter its judgment is far superior to that of the average of individuals, and superior probably in many cases to that which even the best individuals could form for themselves.

How does public opinion come to be superior to individual and to average opinion? There seems to be something paradoxical in the statement.

The fact is of the utmost importance; for public opinion is the ultimate source of sanctions of all public acts, the highest court of appeal before which every executive act performed in the name of the nation must justify itself. If public opinion were merely the immediate expression of the collective feelings and judgments of an unorganised mass of men, its verdicts would be (as we have seen) inferior to those of the average individuals; whereas, as a matter of fact, its expressions are much superior to those of the average individuals.

The influence of public opinion is especially clear and interesting in its relations to law. In this country it is not made by law, but makes law. Where law is imposed and long maintained by the authority of despotic power, it will of course mould public opinion; but, in any progressive highly organised nation, law and the lawyers are always one or two or more generations behind public opinion. The most progressive body of law formally embodies the public opinion of the past generations rather than of the generation living at the time.

The fact of the superiority of public opinion is generally admitted and various explanations are current, for the most part very vague and incomplete. There is the mystical explanation embodied in the dictum that the voice of the people is the voice of God. A rather less vague explanation is that adopted by Mr. Beattie Crozier[1] (among others). It is said that the average man carries within him a germ of an ideal of justice and right, and that he applies this to the criticism or approval of the actions of other men; though he often fails to apply it to his own actions, because, where his own interests are concerned, he is apt to be the sport of purely egoistic impulses.

But this explanation is only partially true. It represents the average man as more hypocritical than he really is, and as falling farther below the standards he acknowledges than he actually does fall. It leaves unexplained the fact that he has this sentiment for an ideal of justice and right; and it proceeds on a false assumption as to the nature of the problem, in assuming that men judge the actions of other men by higher standards than those which they apply to their own conduct; whereas this is by no means generally true.

[1] Cp. his *Civilisation and Progress.*

Is it, then, that superior abilities, which enable a man to gain prestige and to impress his ideas and sentiments upon his fellow men and so to influence public opinion, are commonly combined with a natural superiority of moral sentiment, with a love of right and a hatred of injustice? There may be some degree of such natural correlation of superior abilities with superior moral qualities, but the supposition seems very doubtful; and certainly, if it exists, it is not sufficient to account for the elevation of public opinion. We frequently see consummate ability combined with most questionable moral sentiments, as in Napoleon and many other historic personages.

The true explanation is, I submit, to be found in the basal fact that the moral sentiments are essentially altruistic, while the immoral and non-moral sentiments are in the main self-regarding.[1] Hence, the person who has great abilities but is lacking in moral sentiments and altruism applies his abilities to secure his personal satisfactions and aggrandisement; and, in so far as he aims at affecting the minds of others, he tries only to secure their obedience to his commands and suggestions, to inspire them with deference, admiration, fear and awe, and to evoke an outward display of these feelings. But, as to the ideas and sentiments of the people in general, save in so far as they affect his own gratification, he cares nothing. Accordingly we never find great abilities deliberately, consistently, and directly applied to the degradation of public opinion and morals, save occasionally in relation to some particular end. And we find few or no great works of literature and art deliberately aiming at such degradation.

But with those persons in whom great abilities are naturally combined with moral disposition the case is very different. The moral disposition is essentially altruistic;

[1] On the nature and development of the moral sentiments in the individual mind, see my *Social Psychology*, Chapter VIII.

it is concerned for the welfare of others, of men in general. Hence such a man deliberately applies his abilities to influence the minds of others. The exertion of such influence is for him an end in itself. He seeks and finds his chief satisfaction in exerting an influence, as wide and deep as possible, over the minds of men; not merely in evoking fear or admiration of himself, but in inspiring in them the same elevated sentiments and sympathies which he finds within himself.

For this reason such men as G. F. Watts, Carlyle, and Ruskin exert a much greater and more widespread and lasting influence over the minds of men than do equally able men who are devoid of moral disposition; for the former make the exertion of this influence their chief end, while the others care not at all about the state of public opinion and the minds of the mass. Still less does the non-moral man of great ability strive with all his powers to make others act upon base motives like his own and to degrade their sentiments; rather, he sees that he can better accomplish his selfish ends if other men are unlike himself and are governed by altruistic sentiments; and he sees also that he can better attain his ends if he does lip-service to altruistic ideals; and he is, therefore, apt to exert whatever direct influence he has over the sentiments of men in the same direction as the moral leaders, praising the same actions, upholding in words the same ideals. In this way the men of great abilities, but of immoral or non-moral character, actually aid the moral leaders to some extent in their work; whereas under no conditions is the relation reversed; the moral leaders never praise or acquiesce in bad actions, but always denounce them and use their influence against them.

It follows that, in a well-organised nation, public opinion, which is formed and maintained so largely by the influence of leading personalities, will usually be more in conformity with the sentiments of the best men than of the average man, will be above rather than below private opinion. For, if the bad and the good men of exceptional powers were equal in numbers and capacity, the sum of their influences tending directly to exalt public opinion would be enormously greater than the sum of their influences tending to degrade it; and, as a matter of fact, the influence for good of a few altruistic leaders is able to outweigh the degrading influences of a much larger number of purely selfish men of equally great capacities, and is able to maintain a high standard of public opinion.

We have distinguished a formal and an informal organisation of the national deliberative processes, the latter expressing itself as public opinion. These two organisations co-exist and are, of course, not altogether independent of one another; yet they may be to a considerable extent independent; though the more intimate the functional relations and the greater the harmony between them the healthier will be the national life.

We may note in passing an interesting difference in respect to organisation of the national mind between the English and the American peoples, a difference which illustrates this relative independence of the formal and informal organisations.

In England both the formal and informal organisations have achieved a pretty good level; in both cases the best minds are enabled to exert and have long exerted a dominant influence; and the interaction between the two organisations is very intimate. But in America, while the informal organisation expressed in public opinion seems to be very highly developed, the formal organisation is much inferior; it has not yet such traditions as give the greatest influence to the best minds and embody the effects of their influence. And the better Americans tend to value lightly the formal organisation, to take no part in the working of it, deliberately to ignore it, and to rely rather upon public opinion to repress any evils when they are in danger of reaching an intolerable development.

Both in the formal organisation of the national mind, which is the parliamentary or other national assembly, and in the informal organisation which is public opinion, we see, then, that (in the nation of higher civilisation at least) organisation results in a raising of the collective mental process above the level of the average minds, because it gives a predominant influence to the best minds who form and maintain the traditions, especially the moral traditions; and these press upon the minds of all members of the community from their earliest years, moulding them more or less into conformity with themselves, fostering the better, repressing the purely egoistic, tendencies.

And the ideal organisation after which we ought to strive, is that which would give the greatest possible influence of this sort to the best minds, an influence which consists not in merely organising and directing the energies of the people in the manner most effective for material or even scientific progress, as in modern Germany; but one which, by moulding the sentiments and guiding the reasoning of the people in all matters, public and private alike, secures their consent and agreement and the co-operation of their wills in all affairs of national importance.

When such organisation is in any degree attained and a more or less consistent system of national traditions is embodied in the political, religious, literary, and scientific culture, which moulds in some degree the minds of all men, the national mind clearly becomes, as we said in an earlier chapter, a system of interacting mental forces which are not merely tendencies of the living members of the nation, but are also, in an even greater degree, the ideas and tendencies of the dead; and we see also that in such a people the national consciousness is most truly embodied, not in the minds of the average men, but in the minds of the best men of the time.

The term "public opinion" is sometimes, perhaps generally, used in a looser and wider sense than the meaning implied in the foregoing pages. It is used in the looser sense by President Lowell in his *Public Opinion and Popular Government.* By "public opinion" he seems to mean simply the algebraic sum or balance of individual opinions; he writes "the opinion of the whole people is only the collected opinions of all the persons therein."[1] In accordance with this view, he regards representative institutions as merely one means by which this sum of opinions may be collected and recorded. And he seems to be prepared to regard the "referendum" or the "initiative" in any of their forms, or other methods of direct legislation, as equally good methods, if only all individuals would take the trouble to register their votes upon every question proposed to them. He is aware, of course, that this can hardly be expected of persons who have other interests and occupations than the purely political, and that the direct methods are therefore impracticable as general methods of legislation. If it were true that representative institutions do and should merely collect and record the individual opinions of all members of the public, then it is obvious that each representative should be merely a delegate sent to record the votes of the majority of his constituents. Whereas, if representative institutions should, and in various degrees do, constitute the formal deliberative organisation of the national mind, through which national deliberation and judgment are raised to a higher plane than that of a mere crowd, it follows that the representative should exert his own powers of reasoning and judgment, aided by his special knowledge and equipment, by the special sources of information that he enjoys, in the light

[1] *Op. cit.*, p. 210.

of the discussions in which he takes part, and influenced by all those political traditions whose force he experiences in exceptional fulness by reason of his privileged position. President Lowell, in discussing the functions of the representative, does not decide in favour of the former view, as consistency should perhaps lead him to do; thereby showing that he is not wholly committed to the individualist view. He discusses the question whether the member of Parliament or Congress should regard himself as representing the interests of his constituents alone, or as concerned primarily and chiefly with the interests of the whole people; and he rightly inclines to the latter view. This is not quite the same distinction as that which is insisted upon in these pages. Even if each representative were concerned only for the welfare of the nation as a whole, yet so long as he regarded it as his sole function to vote as he believes the majority of the citizens would vote in any process of direct legislation, he would fall short of the highest duty which is laid upon him by his position—namely, not merely that of recording the opinion of the majority, but that of taking part in the organised deliberative activities of the national mind by which it arrives at judgments and decisions of a higher order than any purely individual, or algebraic sum of individual, judgments and decisions.[1]

Public opinion, in the sense in which I have used the words in this chapter (which seems to me the only proper use of them) is, then, not a mere sum of individual opinions upon any particular question; it is rather the expression of that tone or attitude of mind which prevails throughout the nation and owes its quality far more to the influence of the dead than of the living, being the expression of the moral sentiments that are firmly and traditionally estab-

[1] In this connection I would refer the reader to *The New State* by M. P. Follett (London, 1918) an interesting book in which the true nature and function of collective deliberation are forcibly expounded.

lished in the mind of the people, and established more effectively and in more refined forms in the minds of the leaders of public opinion than in the average citizen. This tone of the national mind enables it to arrive at just judgments on questions of right and wrong, of duty and honour and public desert; though it may have little bearing upon such practical questions as bimetallism, tariff reform, or railway legislation. The current use of the term, in this country at least, does, I think, recognise that public opinion properly applies only to the sphere of moral judgments and can and should have no bearing upon the practical details of legislation. Public opinion is, both in its development and in its operations, essentially collective; it is essentially the work of the group mind. Its accepted standards of value are slowly built up under the influence of the moral leaders of past ages; and, in the application of those standards to any particular question, the influence of the moral leaders of the time makes itself felt. I have kept in mind in the foregoing pages the public opinion of the nation; but every community, every association, every enduring group has its own public opinion, which, though it is influenced by, and indeed is, as it were, a branch of, the main stem of national public opinion and is therefore of the same fibre and texture, has nevertheless its own peculiar tone and quality, especially in regard to the moral questions with which each group is specially concerned.

PART III

THE DEVELOPMENT OF NATIONAL MIND AND CHARACTER

CHAPTER XIV

Introductory

IN the first Part of this book we have reviewed the most general principles of collective mental life, beginning with the unorganised crowd as affording the simplest example, considering then an army as the simplest example of the profound modifications of collective mental life effected by organisation of the group. In the second Part we passed on to apply these principles to the understanding of the mind of the nation as the most important, complex, and interesting of all types of the group mind.

In the third Part I take up the consideration in a general way of the processes by which national mind and character are gradually built up and shaped in the long course of ages. For just as we cannot understand individual minds, their peculiarities and differences, without studying their development, so we cannot hope to understand national mind and character and the peculiarities and differences of nations, without studying the slow processes through which they have been built up in the course of centuries.

In an earlier chapter, in connection with the question of the importance of homogeneity of mental qualities as a condition of the existence of the national mind, I argued that race has really considerable influence in moulding the type of national mind. I recognised that differences of innate qualities between races, at any rate between allied subraces, are not great, and that they can be, and generally are, almost completely over-ridden and obscured in each individual by the moulding power of the social environment in which he grows up; but I urged that these racial qualities are very persistent, and that they exert a slight but constant pressure or bias upon the development of all that constitutes social environment, upon the forms of institutions, customs, traditions, and beliefs of every kind, so that the effect of such slight but constant bias accumulates from generation to generation, and in the long run exerts an immense influence.

One way of treating the part played by the racial mental qualities in the development of the national mind would be to attempt to define the racial or innate peculiarities of the peoples existing at the present time, and to assume that these peculiarites were produced in the remote past, before the formation of nations began, and that they have persisted unchanged throughout the period of the development of nations. Something of this sort was proposed by Walter Bagehot in his *Physics and Politics*. He distinguished in the development of peoples two great periods—on the one hand the race-making period, which roughly corresponds to the whole prehistoric period, and on the other hand the nation-making period, which roughly corresponds to the historic period. This distinction has undoubtedly a certain validity.

It seems probable that man was evolved from his prehuman ancestry as a single stock, probably a stock somewhat widely distributed in the heart of the Eurasian continent, or possibly in Africa according to the recent view of some authors, or in the area which is now the Indian Ocean. If this be true, it follows that the differentiation of the mental and physical qualities of the principal human races, the differentiation of the white and black and yellow and brown races, as well as of the chief subraces, such as the Semitic, the races of Europe—the *Homo*

Europæus, *Alpinus* and *Mediterraneus*—was the work of the immensely prolonged prehistoric period. For these races and subraces, as we now know them, seem to have been in existence and to have had recognisably and substantially the same leading qualities, both mental and physical, that they now have, before the beginning of the historic period.

The racial differentiation during the prehistoric period must have been much greater than during the historic period; and this was not only because the former period was immensely longer, but also because, in all probability, the rate of racial change has been on the whole slower in the historic period.

The differentiation of racial types in the prehistoric period must have been in the main the work of differences of physical environment, operating directly by way of selection, by way of the adaptation of each race to its environment through the extermination of the strains least suited to exist under those physical conditions. But this process, this direct moulding of racial types by physical environment, must have been well nigh arrested as soon as nations began to form. For the formation of nations implies the beginning of civilisation; and civilisation very largely consists in the capacity of a people to subdue their physical environment, or at least to adapt the physical environment to men's needs to a degree that renders them far less the sport of it than was primitive man; it consists, in short, in replacing man's natural environment by an artificial environment largely of his own choice and creation.

In a second and perhaps even more important way, the formation of nations, with the development of civilisation, modified and weakened the moulding influence of the physical environment; namely, it introduced social cooperation in an ever increasing degree, so that the perpetual struggle of individuals and of small family groups with one another and with nature was replaced by a co-operative struggle of large communities against the physical environment and with one another. And in this process those members of each community who, by reason of weakness, general incapacity, or other peculiarity, would have been liable to be eliminated under primitive conditions became shielded in an ever increasing degree by the powers of the stronger and more capable against the selective power of nature and against individual human forces. And, although within the community the rivalry of individuals and families still went on, it was no longer so much a direct struggle for existence, but rather became more and more a struggle for position in the social scale; and failure in the struggle no longer necessarily meant death, or even incapacity to leave an average number of descendants. That is to say, primitive man's struggle for existence against the forces of nature and against his fellow men, which made for racial evolution and differentiation through survival of those fittest to cope with various environments, tended to be replaced by a struggle which no longer made for racial evolution towards a higher type, and which may even have made for race-deterioration, at the same time that civilisation and national organisation continued to progress.

We may, then, recognise a certain truth in Bagehot's distinction of two great periods, the race-making and the nation-making periods. Neverthless, it would not be satisfactory to follow the course suggested above and simply assume certain racial characters as given fixed data without further consideration. For, firstly, it is interesting and perhaps not altogether unprofitable to indulge in speculations on the race-making processes of the prehistoric period. Secondly, although it seems likely that racial changes have been in the main slower and on the whole relatively slight in the historic period, yet they have not been altogether lacking; and, in proportion to their magnitude, such changes as have occurred have been of great importance for national life; and changes of this kind are still playing their part in shaping the destinies of nations. Possible racial changes of mental qualities must therefore be considered, when we seek to give a general account of the conditions of the development of nations.

On the other hand, we must reject root and branch the crude idea, which has a certain popular currency, that the development of civilisation and of nations implies a parallel evolution of individual minds. That idea we have already touched upon and rejected in a previous chapter, where we arrived at the conclusion that there is no reason to suppose the present civilised peoples to be on the whole innately superior to their barbaric ancestors.

If we use the word "tradition" in the widest possible sense to denote all the intellectual and moral gains of past generations, in so far as they are not innate but are handed on from one generation to another by the personal intercourse of the younger with the older generation, and if we allow the notion of tradition to include all the institutions and customs that are passed on from generation to generation, then we may class all the changes of a people that constitute the evolution of a national character under the two heads: *evolution of innate qualities* and *evolution of traditions*. Using the word "tradition" in the wide sense just now indicated, the traditions of a people may be said to include the recognised social organisation of the whole people into classes, castes, clans, phratries, or groups of any kind, whose relations to one another and whose place in the national system are determined by law, custom, and conventions of various kinds. This part of the total tradition is relatively independent of the rest, and we may usefully distinguish the development of such social organisation as social evolution—giving to the term this restricted and definite meaning—and we may set it alongside the other two conceptions as of co-ordinate value.

If we thus set apart for consideration under a distinct head the evolution of social organisation, the rest of the body of national traditions may be said to constitute the civilisation of a people. For the civilisation of a people at any time is essentially the sum of the moral and intellectual traditions that are living and operative among them at that particular time. We are apt in a loose way to consider the civilisation of a people to consist in its material evidences; but it is only in so far as these material evidences, the buildings, industries, arts, products, machinery, and so forth, are the expression and outcome of its mental state that they are in any degree a measure of its civilisation. We may realise this most clearly by considering the case of a people on which the material products of civilisation have been impressed from without. Thus the peasants of India live amongst, and make use of, and benefit materially by, the railways and irrigation works created by their British rulers, and are protected from invasion and from internal anarchy by the British military organisation and equipment; and they play a subordinate though essential part in the creation and maintenance of all these material evidences of civilisation. But these material evidences are not the expression of the mental state of the peoples of India, and form no true part of their civilisation; and, in fact, they affect their civilisation astonishingly little; although if these products of a higher civilisation should be maintained for a long period of time they would, no doubt, produce changes of their civilisation, probably tending in some degree to assimilate their mental state to that of Western Europe.

We may, then, with advantage distinguish between the social organisation and the civilisation of a people. In doing so we are of course making an effort of abstraction which, though it results in an artificial separation of things intimately related, is nevertheless useful and therefore justifiable. In a similar way the progress of civilisation may be distinguished from social evolution. Social evolution is profoundly affected by the progress of civilisation, and in turn reacts powerfully upon it; for any given social organisation may greatly favour or obstruct the further progress of civilisation. There could have been no considerable advance of civilisation without the evolution of some social organisation; but that the two things are distinct is clear, when we reflect that there may be a very complex social organisation, implying a long course of social evolution, among a people that has hardly the rudiments of civilisation. Extreme instances of social organisation in the absence of civilisation are afforded by some animal societies—for example, societies of ants, bees, and wasps. Among peoples, the native tribes of Australia illustrate the fact most forcibly. They are at the very bottom of the scale of civilisation; yet it has been discovered that they have a complex and well-defined social organisation, which can only have been achieved by a long course of social evolution. These people are divided into totem clans, which clans are grouped in phratries, each individual being born, according to well-recognised rules, into a clan of which he remains a life-long member; and his membership in the clan and phratry involves certain well-defined rights and obligations, and well-defined relations to other persons, especially as regards marriage; and these rights, obligations, and relations are recognised and rigidly maintained throughout immense areas.

On the other hand, although no people has attained any considerable degree of civilisation without considerable social organisation, nevertheless we can at least imagine a people continuing to enjoy a high civilisation, practising and enjoying much of the arts, sciences, philosophy, and literature, which we regard as the essentials of civilisation, yet retaining a bare minimum of social organisation. And this state of affairs is not only conceivable, but is held up as a practicable ideal by philosophical anarchists such as Tolstoi and Kropotkin; and it is, I think, true to say that the American nation presents an approximation to this condition.

Again, a very high state of civilisation may co-exist with a relatively primitive social organisation. Thus the civilisation of Athens in the classical age was equal to, or even superior to, our own in many respects; yet the social organisation was very much less highly evolved. It had hardly emerged from the barbaric patriarchal condition, and had at its foundation a cruel system of slavery;[1] and it had also another great point of inferiority—namely, the very restricted number of persons included in the social system. These deficiencies, this rudimentary character, of its social organisation was the principal cause of the instability and brief endurance of that brilliant civilisation.

We have so far distinguished three principal factors or groups of factors in the evolution of national mind and character: (1) Evolution of innate or racial qualities: (2) Development of civilisation: (3) Social evolution, or the development of social organisation.

Now the first two of these we may with advantage divide under two parallel heads, the heads of intellectual and moral development. No doubt, the intellectual and the moral endowment of a people continually react on each other; and many of the manifestations of the national mind are jointly determined by the intelligence and the morality of a people; especially perhaps is this true of their religion and their art. Nevertheless, it is clear that we can distinguish pretty sharply between the intellectual and the

[1] Cf. W. R. Patterson, *The Nemesis of Nations*, London, 1906.

moral traditions of a people; and that these may vary independently of one another to a great extent. A rich and full intellectual tradition may go with a moral tradition of very low level, as in the Italian civilisation of the renascence; and a very high moral tradition with a relative poverty of the intellectual, as in the early days of the puritan settlements of New England.

The same distinction between the intellectual and the moral level is harder to draw in the case of the racial qualities of a people, but it undoubtedly exists and is valid in principle, no matter how difficult in practice to deal with.

We have, then, to distinguish five classes of factors, five heads under which all the factors which determine the evolution of national character may be distributed. They are

(1) Innate moral disposition
(2) Innate intellectual capacities } racial qualities.
(3) Moral tradition
(4) Intellectual tradition } national civilisation.
(5) Social organisation.

Every nation that has advanced from a low level to a higher level of national life has done so in virtue of development or progress in one or more of these respects. And a principal part of our task, in considering the evolution of national mind and character, is to assign to each of these its due importance and its proper place in the whole complex development.

The distinction between the racial and the traditional level of a people is too often ignored; chiefly, perhaps, for the reason that it has usually been assumed that whatever is traditional becomes innate and racial through use. Since in recent years it has been shown that this assumption is very questionable, a number of authors have recognised the importance of the distinction as regards the intellectual qualities of a people; but, as regards the moral qualities, the distinction is still very generally overlooked.

The neglect of these distinctions between the innate and the traditional has in great measure vitiated much of the keen dispute that has been waged over the question whether the progress of civilisation depends primarily on intellectual or on moral advance. For example, T. H. Buckle and Benjamin Kidd agreed in recognising clearly the distinction between the innate and the traditional intellectual status of a people; and they agreed in maintaining that we have no reason to believe that in the historic period any people has made any considerable advance in innate intellectual capacity; and that any such advance, if there has been any, has not been a principal factor in the progress of civilisation. But they differed extremely in that Buckle maintained that the primary cause of all progress of national life is the improvement of its intellectual tradition, that is, increase in the quantity and the worth of its stock of knowledge and accepted beliefs, and improvements in methods of intellectual operation; and he held that improvements of morals and of social organisation have been secondary results of these intellectual gains. Kidd, on the other hand,[1] maintained that the progress of European civilisation has been primarily due to an improvement of the morality of peoples; that this has led to improvement of social organisation; and that this in turn has been the essential condition of the progress of the intellectual tradition, because it has secured a stable social environment, a security of life, a free field for the exercise of intellectual powers; in the absence of which conditions the intellectual powers of a nation cannot effectively or-

[1] *Social Evolution*, *Principle of Western Civilisation*, and *The Science of Power*.

ganise themselves and apply themselves to the understanding of man and nature, or to securing the traditional perpetuation of the gains which they may sporadically achieve. We have to examine these views and try to determine what truth they contain, and to show that they are not wholly opposed but can in some measure be combined.

I propose to make first a very brief critical survey of some of the most notable attempts that have been made to account for racial qualities, and I shall try to supplement and harmonise these as far as possible. We may with advantage consider at the outset the race-making period, and afterwards go on to consider changes of racial qualities in the historic period. This Part of the book is necessarily somewhat speculative, but its interest and importance for our main topic may justify its inclusion.

CHAPTER XV

The Race-Making Period

LET us now see what can be said about the process of racial differentiation which, as we saw in the foregoing chapter, was in its main features accomplished in the prehistoric or race-making period. We cannot hope to reach many positive conclusions, but rather merely to discuss certain possibilities and probabilities in regard to the main factors of the differentiation of racial mental types.

I would point out at once that the answer to be given to the question—Are acquired qualities transmitted? Are the effects of use inherited? is all important for our topic. I do not propose to discuss that difficult question now. I will merely say that the present state of biological science makes it seem doubtful whether such inheritance takes place, and that, although the question remains open, we are not justified in assuming an affirmative answer; that, therefore, we must not be satisfied with any explanation of racial and national characteristics based upon this assumption; and in the following discussion I shall provisionally assume the truth of the Neo-Darwinian principle that acquired modifications are not transmitted.

Assuming, as we must, that all peoples are descended from some one original stock, the problem is—Can anything be said of the conditions which have determined the differentiation of races of different mental constitutions, of the development of racial qualities which, having become relatively fixed, have led to the evolution of different types of national organisation and culture? And especially we have to consider the conditions which have produced, and may still produce in the future, the qualities that make for the progress of nations.

We must suppose a certain social organisation to have obtained among that primitive human stock from which all races have been evolved, probably an organisation in small groups based on the family under the rule and leadership of a patriarch.

It is possible that considerable divergences of social organisation may have taken place, without any advance towards civilisation; such divergences of social organisation must have tended to divert the course of mental evolution along various lines; but they must themselves have had their causes; they cannot in themselves be the ultimate causes of divergence of racial mental types.

Such ultimate causes of the differentiation of mental qualities must have been of two orders only, so far as I can see: (1) differences of physical environment: (2) spontaneous variations in different directions of the innate mental qualities of individuals, especially of the more gifted and energetic individuals of each people.

In the mental evolution of animals these two factors are not distinguishable. We may say that the main and perhaps the sole condition of their evolution is the selection by the physical environment of spontaneous favourable variations and mutations of innate mental qualities; if we include under the term physical environment of the species all the other animal and vegetable species of its habitat. For it is only by its selective influence upon individual variations that physical environment can determine differentiation of races.

But with man the case is different; spontaneous variation not only provides the new qualities which, by determining the survival of the individual in his struggle for existence with the physical environment, secure their own perpetuation by transmission to the aftercoming generations. The new qualities determine mental evolution in another manner, by a mode of operation which is almost completely absent in animal evolution; namely, the spontaneous variations create a social environment which profoundly modifies the influence of the physical environment, and itself becomes a principal factor in the determination of the trend of racial evolution.

Man is distinguished from the animals above all things by his power of learning. Whereas the behaviour of animals, even of the higher ones, consists almost entirely of purely instinctive actions, innate modes of response to a limited number of situations; man has an indefinitely great capacity for acquiring new modes of response, and so of adapting himself in new and more complex ways to an almost indefinite variety of situations. And his new mental acquisitions are not made only by the slow process of adaptation in the light of his own individual experience of the consequences of behaviour of this and that kind; as are most of the few acquisitions of the animals. By far the greater part of the mental stock-in-trade by which his behaviour is guided is acquired from his fellow men; it represents the accumulated experience of all the foregoing generations of his race and nation. Man's life in society, together with the great plasticity of his mind, its great capacity for new adaptations, secures him this enormous advantage; the two things are necessarily correlated. Without the plasticity of mind, his life in society would benefit him relatively little. Many animals that lead a social life in large herds or flocks are not superior, but rather inferior, in mental power to animals that lead a more solitary life; and indeed this seems to be generally true, as we see on comparing generally the herbivorous gregarious animals with the solitary carnivores that prey upon them. The social life of such animals, rendering individual intelligence less necessary for protection and escape from danger, tends actually against mental development.

On the other hand, man's great plastic brain would be of comparatively little use to him if he lived a solitary unsocial life. His great brain is there to enable him to assimilate and make use of the accumulated experience, the sum of knowledge and morality, which is traditional in the society into which he is born a member; that is to say, the development of social life, which depended so much upon language and for the forwarding of which language came into existence, must have gone hand in hand with the development of the great brain, which enables full advantage to be secured from social co-operation and which, especially, renders possible the accumulation of knowledge, belief, and traditional sentiment.

Now this traditional stock of knowledge and morality has been very slowly accumulated, bit by bit; and every bit, every least new addition to it, has been a difficult acquisition, due in the first instance to some spontaneous variation of some individual's mental structure from the ancestral type of mental structure. That is to say, throughout the evolution of civilisation, progress of every kind, increase of knowledge or improvement of morality, has been due to the birth of more or less exceptional individuals, individuals varying ever so slightly from the ancestral type and capable, owing to this variation, of making some new and original adaptation of action, or of perceiving some previously undiscovered relation between things.

These new acqusitions, first made by individuals, are, if true or useful, sooner or later imitated or accepted by the society of which the original-minded individual is a member, and then, becoming incorporated in the traditional stock of knowledge and morality, are thereby placed at the service of all members of that society.

Thus, favourable spontaneous variations do not, as with the animals, render possible mental evolution merely by conducing to the survival of, and the perpetuation of the qualities of, those individuals in whom the variations occur. They may do this, or they may not; but, in addition and more importantly, they contribute to the stock of traditional knowledge and morality, and so raise the social group as a whole in the scale of civilisation; they render it more capable of successfully contending against other groups and against the adverse influence of the physical environment; and they promote the solidarity of the group by adding to its stock of common tradition; thus the acquisitions of each member benefit the group as a whole and all its members, quite apart from any philanthropic purpose or intention of producing such a result.

The achievement of this unconscious undesigned solidarity of human societies is one of two great steps in the evolution of the human race by which the process is rendered very different from, and is raised to a higher plane than, the mental evolution of the animal world. The second and still more important step is one which is only just beginning to be achieved in the present age; I shall have to touch on it in a later chapter.

The original or primary divergence of mental type between any two peoples must, then, have been due to these fundamental causes—namely, differences of physical environment and spontaneous variations of mental structure, the latter adding to the traditional stock of knowledge and belief, of moral precepts and sentiments.

Intellectual or moral divergence produced by these two primary causes would tend to determine the course of social evolution along different lines and so to produce different types of social organisation. And different social organisations thus produced would then react upon the moral and intellectual life of the people to produce further divergence; for example, one type of social organisation determined by physical environment, say a well developed patriarchal system, may have made for progress of intellect and morals; another, say a matriarchal organisation, or one based on communal marriage, may have tended to produce stagnation.

As social evolution proceeded and brought about more extensive and more complex forms of social organisation, which included, within any one society or group, larger numbers of individuals in more effective forms of association, social organisation must have assumed a constantly increasing importance as a condition of mental evolution relatively to all other factors, especially as compared with the influence of physical environment; until, in the complex societies of the present time, it has an altogether predominant importance. This truth is concisely stated in the old dictum that "in the infancy of nations men shape the State; in their maturity the State shapes the men." Accordingly, in considering the mental evolution of peoples we must never lose sight of the influence of social organisation. It follows that the conditions of the mental evolution of man are immensely more complex than those of the mental evolution of animals.

We must recognise not only the selection, through survival in the struggle for existence, of new mental qualities arising as spontaneous variations of individual mental structure. This, which is the only, or almost the only, process at work in the mental evolution of animals, is immensely complicated and overshadowed in importance by two processes. The first is the accumulation of knowledge and morality in traditional forms. The traditional accumulation, which so far outweighs the mental equipment possible to any individual isolated from an old society, not only constitutes in itself a most important evolutionary product, but it modifies profoundly the conditions of evolution of the individual innate qualities of mind; for example, the greater and more valuable the stock of traditional knowledge and morality becomes, the more does fitness to survive consist in the capacity to assimilate this knowledge and to conform to these higher moral precepts, the less does it consist in the purely individualistic qualities, such as quickness of eye and ear, fleetness of foot, or strength and skill of hand. Secondly, the processes of natural selection are complicated by the social evolution, which tends progressively to abolish the struggle for existence between individuals, and to replace it by a struggle between groups; in which struggle success is determined not only by the qualities of individuals, but also very largely by the social organisation and by the traditional knowledge and morality of the groups.

Each variety of the human species, each race considered as a succession of individuals having certain innate mental qualities, has been evolved, then, not merely under the influence of the physical environment, like the animal species, but also and to an ever increasing extent under the influence of the social environment. The social environment we regard as consisting of two parts; namely, the social organisation and the body of social tradition; for these, though interdependent and constantly interacting may yet with advantage be kept apart in thought. We must, then, bear constantly in mind the fact that man creates for himself an environment which becomes ever more complex and influential, overshadowing more and more in importance the physical environment.

Here I would revert to some points of the analogy, drawn in Chapter X, between the mind of a nation and that of an individual. The mind of an individual human being develops by accumulating the results of his experience; and so does that of a people. In this respect the analogy holds good. But the development possible to an individual is strictly limited in two ways. First, by the short duration of the material basis of his mental life; secondly by the extent of his innate capacities. Neither of these limitations applies to the national mind. Its material basis is in principle immortal, because its individual components may be incessantly renewed; and its development has no limit set to it by its innate capacities, because these may be indefinitely extended and improved. In these respects the national mind resembles the species rather than the individual.

The development of the national mind, and of the minds of those who share in the mental life of the nation, thus combines the methods and advantages of the development of individuals and of species, methods which are essentially different. The result is that the mental development of man, since his social life began, has been radically different from that of the animals; it has been a social process; it has been the evolution of peoples rather than of individuals. The evolution of man as an individual has been subordinated to that of peoples; and it is incapable of being understood or profitably considered apart from the development of the group mind.

Assuming, as we must, that all the races of men are derived from a common stock, it is obvious, I think, that the first differentiation of racial types was determined almost exclusively by differences of physical environment; and that the other conditions only very slowly developed and did not assume their predominant importance until the time which may be roughly defined as the beginning of the historic or nation-making period.

Physical environment affects the mental qualities of a people in three ways: firstly, it directly influences the minds of each generation; secondly, it moulds the mental constitution by natural selection, adapting the race to itself; thirdly, it exerts indirect influence by determining the occupations and modes of life and, through these, the social organisation of a people. We may consider these three modes of influence in turn.

There has been much speculation on the direct influence of the physical environment in moulding the mental type of a people; but little or nothing can be said to be established.

There is a fair concensus of opinion to the effect that what we may call climate exerts an important influence. In climate the two factors recognised as of chief importance are temperature and moisture. High temperature combined with moisture certainly tends to depress the vital activity of Europeans and to render them indolent, indisposed to exertion of any kind. On the other hand, high temperature combined with dryness of the atmosphere seems to have the effect of rendering men but little disposed to continuous activity, and yet capable ot great efforts; it tends to produce a violent spasmodic activity. A cold climate seems to dispose towards sustained activity and, when combined with much moisture, to a certain slowness.

These effects, which we ourselves experience and which we see produced upon other individuals on passing from one climate to another, we seem to see impressed upon many of the races which have long been subjected to these climates; for example, the slow and lazy Malays have long occupied the hottest moistest region of the earth. The Arabs and the fiery Sikhs may be held to illustrate the effect of dry heat. The Englishman and the Dutchman seem to show the effects of a moist cool climate, a certain sluggishness combined with great energy and perseverance.

In these and other cases, in which the innate temperament of a people corresponds to the effects directly induced by their climate, it seems natural to suppose that the innate temperament has been produced by the transmission and accumulation from generation to generation of the direct effects of the climate. The assumption is so natural that it has been made by almost every writer who has dealt with the question. And these instances of conformity of the temperament of peoples to the direct effects of climate are sometimes offered as being among the most striking evidences of the reality of hereditary transmission of acquired qualities; and the argument is reinforced by instances of what seem to be similar results produced by climate on physical types. Thus, it is said that in North America a race characterised by a new specific combination of mental and physical qualities is being rapidly formed; and it seems to be well established that long slender hands are among these features; for in Paris a specially long slender glove is made every year in large quantities for the American market. Again, we see apparently a change of physical type in the white inhabitants of Australia. They seem to be becoming taller and more slender "cornstalks;" and this is commonly regarded as the direct effect of climate.

Now, that a new race or subrace with a specific combination of qualities should be forming in America is certainly to be expected from the fact that the intimate blending of a number of European stocks has been going on for some generations. But what gives special support to the assumption that these new qualities are the direct effects of climate is that these qualities, the physical at least, seem to be approximations to the type of the Red Indians, the aboriginal inhabitants. And, it is said, this approximation of type can be due only to hereditary accumulation of the direct effects of the climate on individuals.

Another way in which climate has been held to modify

racial mental qualities by direct action is through the senses, especially the eye. M. Boutmy, in his book on the English people makes great play with this principle.[1] He points out that the thick hazy state of the air, so common in our islands, renders vague and dull all outlines and colours, so that the eye does not receive that wealth of well-defined hues and forms which give so great a charm to some more sunny lands, such as the Mediterranean coast lands. Hence, he says, the senses become or remain relatively dull, and the sense-perceptions slow and relatively indiscriminating. Such relative deficiency of æsthetic variety and richness in the appearance of the outer world produces secondarily a further and deeper modification of mental type. In the lands where nature surrounds man with an endless variety of rich and pleasing scenes, he can find sufficient satisfaction in mere contemplation of the outer world; and, when he takes to art production, he tends merely to reproduce in more or less idealised forms the objects and scenes he finds around him; his art tends to be essentially objective. On the other hand, in the dull northern climes, man has not ever at hand these sources of satisfaction in the mere contemplation of the outer world; consequently he is driven back upon his own nature, to find his satisfactions in a ceaseless activity of mind and body, but chiefly of the latter. Hence, races so situated are characterised by great bodily activity and their art and literature are essentially subjective. The thick air, the monotony of vague form and colour, drive the mind to reflection upon itself; and in art the objects of nature serve merely as symbols by aid of which the mind seeks to express its own broodings. "The painter paints with the intentions of the poet, the poet describes or sings with the motives of the psychologist or moralist. All the literature of imagination of the English

[1] *Essai d'une Psychologie politique du Peuple Anglais*, Paris, 1903.

shows us the internal reacting incessantly upon the external with a singular power of transfiguration and interpretation."[1] Hence also poetry is the privilege of a few rare spirits and is for them the product of deep reflection, not a simple lyrical expression in which all can equally share.

It is certainly true that climate tends to produce these effects by its direct action on individuals. Anyone who has lived for a time in the southern climes must have noted these effects upon himself. But we have no proof that the effects of climate are directly inherited. It suffices to suppose that the direct effects are imposed afresh by the climate on the minds of each generation. This view is borne out by the fact that two races may live for many generations in the same climate and yet remain very different in temperament in these respects; for example the Irish climate is very similar to the English, perhaps even more misty and damp; yet the Irish have much more wit and liveliness than the English. And in every case in which adaptation to physical environment has clearly become innate or racial, an explanation can be suggested in terms of selection of spontaneous variations, or of crossing of races. Thus, the approximation of the American people to the type of the aboriginals, if it is actual, and some observers deny it, may well be due to the small infusion of the native blood which has admittedly taken place. It may well be that certain qualities of the Red Indian, for example, the straight dark hair and prominent cheek bones, are what the biologists call "dominant characters" when the Indian is crossed with the European; that is, qualities which always assert themselves in the offspring to the exclusion of the corresponding quality of the other race involved in the cross. If that is so, a very small proportion of Indian blood would suffice to make these

[1] *Op. cit.* p. 20.

features very common throughout the population of America. As an exception to the supposed law of direct hereditary adaptation to climate take the colour of the skin. The black Negroes live in the hot moist regions of Africa; and it has been said that pigmentation is the hereditary effect of a hot moist climate. But there are men of a different race who have long lived in an equally hot and moist climate, but who do not show this effect—namely, tribes in the heart of Borneo, close to the equator, whose skins are hardly darker than the average English skins and less dark than the Southern Europeans'. Take again the indolence of the peoples of moist hot climates and the energy of peoples of colder climates. These certainly seem to be racial qualities; but their distribution is adequately explained by the indirect effect of physical environment exerted by way of natural selection; and these differences of energy afford the best illustration of such indirect action of physical environment in determining racial mental qualities.

Before considering the question further, let us note yet another way in which the physical environment affects men's minds and has been supposed directly to induce certain racial qualities. Buckle pointed out with great force the influence on the mind of what he called the external aspects of nature. He showed that where, as in India and the greater part of Asia, the physical features of a country are planned upon a very large scale; where the mountains are huge, where rivers are of immense length and volume, where plains are of boundless extent, and the sun very hot, there the forces of nature are exerted with an intensity that renders futile the best efforts of man, at any rate of man in a state of low civilisation, to cope with them. In such countries men are exposed to calamities on an enormous scale, great floods, violent storms and deluges of rain, earthquakes, excessive droughts resulting in famine and plague; and they are exposed to the attacks of many dangerous animal species, which are bred by the great heat in the dense and unconquerable forests. These disasters have repeatedly occurred on a scale such that in comparison with them the recent earthquake in California appears a mere trifle. Millions have been destroyed in a few hours in some of the floods of the Yellow River of China.

The magnitude of these objects and the appalling and irresistible character of such devastating forces produce, said Buckle, two principal and closely allied effects upon the mind; they stimulate the imagination to run riot in extravagant and grotesque fancies; at the same time, they discourage any attempt to cope with these great forces and to understand their laws, and thus keep men perpetually in fearful uncertainty as to their fate; for they cannot hope to control it by their own unaided efforts.

Hence, the encouragement of superstition; hence, the dominance of a degrading religion of fear throughout the greater part of such regions; hence, the supremacy of priests and religious orders and the discouragement of scientific reasoning. Hence, in the arts, the literature, and the religion of India, we see a dominant tendency to the grotesque, the enormous, the fearful; we see gods portrayed with many arms, with three eyes and terrible visages. The legends of their heroes contain monstrous details, as that they lived for many thousands or millions of years. "All this," says Buckle, "is but a part of that love of the remote, that straining after the infinite, and that indifference to the present, which characterises every branch of Indian intellect. Not only in literature, but also in religion and in art, this tendency is supreme. To subdue the understanding, and indulge the imagination, is the universal principle. In the principles of their theology, in the character of their gods, and even in the form of their temples, we see how the sublime and threatening aspects of the external world have filled the mind of the people with those images of the grand and the terrible, which they strive to reproduce in a visible form, and to which they owe the leading peculiarities of their national culture."[1]

That these peculiarities of the mental life of such peoples are causally related to those terrible aspects of nature is, I think sufficiently established by Buckle. But if we admit this, there remain two questions: (1) Have these tendencies become innate racial qualities? (2) If so, how have they been rendered innate? Buckle did not raise these questions and offered no opinion in regard to them. But he seems to have assumed that these tendencies have become innate; and there is much to be said for that view. Yet, if that could be shown conclusively, it still would not prove inheritance of these acquired qualities. It may have resulted in some such way as this: the physical environment stimulates the imagination, and it represses the tendency to control imagination and superstition by reason and calm inquiry after causes; acting thus upon successive generations of men, it determines the peculiarities of the religious system and of the art and literature of the people. Individuals in whom the same tendencies are innately strong will then flourish under such a system; whereas those whose innate tendencies are in the direction of reason and scepticism will find the system uncongenial, unfavourable for the exercise of their best power; they will fail to make their mark; they may, as in many instances of European inquirers, actually have lost their lives or their liberty through the religious zeal of those who maintain the traditional system. Thus the social environment, working through long ages, may have constantly determined a certain degree of selection of the

[1] *History of Civilisation*, p. 137.

innate tendencies congenial to it, and a weeding out of the opposed tendencies; until the former have predominated in the race.[1]

We have here a very important principle which we must constantly bear in mind—namely, that not only the physical environment, but also the social environment, may determine the survival of those temperaments and qualities of mind best fitted to thrive in it, and, by handicapping those least fitted to it, may gradually bring the mental qualities of the race into conformity with itself. We shall later see other examples by which this principle is more clearly illustrated.

We conclude that, while physical environment may act powerfully upon the minds of individuals, moulding their acquired qualities in the three ways noticed—namely, influencing the mind through bodily habit, through the senses, and through the imagination—there is no sufficient evidence that the acquired qualities so induced ever become innate or racial characters by direct transmission. In those instances in which the racial qualities approximate to these direct effects of physical environment, it may well be because the physical environment has brought about adaptation of the race by long continued selection of individuals; or because it has determined peculiarities of social environment, which in turn have brought about adaptation of the racial qualities by long continued selection.

[1] In this connection it must be remembered that in Hindu society the man of proved and acknowledged holiness is permitted and encouraged to procreate a large number of children.

CHAPTER XVI

The Race-Making Period (*continued*)

WE considered in our last chapter the principal modes in which physical environment affects the character of a people—namely, (1) influence on temperament exerted chiefly through climate acting upon the bodily functions: (2) influence through the senses, exerting secondary effects upon the higher mental processes: (3) direct influence on the imagination. We concluded that these effects become innate in some degree; though whether they are impressed on the race by direct inheritance, or by processes of direct or of social selection, or in all three ways, remains an open question.

We distinguished, besides these direct modes of influence, two indirect modes by which physical environment affects the minds and character of a people: (1) by its selective action on individuals apart from its influence upon their minds: (2) by determining occupations and social organisation. We may consider them in turn.

It is recognised, as I pointed out above, that the races inhabiting hot moist countries are commonly indolent, while those of the moderately cold and moist climates tend to be extremely active and energetic.

This difference is well brought out by Mr. Meredith Townsend in an essay on the charm of Asia for the Asiatics[1]; and he is speaking not of Asia in general but of Southern Asia. He says Asiatics "will not, under any

[1] *Europe and Asia.*

provocation, burden themselves with a sustained habit of taking trouble. You might as well ask lazzaroni to behave like Prussian officials." After quoting Thiers' description of the immense labours of detailed administration which he supported while minister of State, he says "No Asiatic will do that. . . . One-half the weakness of every Oriental government arises from the impossibility of finding men who will act as M. Thiers did." These races, bred in the tropics, are in fact incurable lotus-eaters, their chief desire is for the afternoon life or, as is commonly said of the Malays throughout the Eastern Archipelago, they are great legswingers; they prefer to undertake no labour more arduous than sitting still swinging their legs. All this, though more or less true of the tropical races in general, is pre-eminently true of those inhabiting regions which are moist as well as hot, the Malays, the Burmese, the Siamese, the Papuans, the Negroes of the African jungle regions.

Such peoples have failed to acquire the energy which leads men to delight in activity for its own sake, not merely because a hot moist climate inclines directly to indolence, but rather because the prime necessities of life are to be had almost without labour; the heat dispenses with the necessity for clothing and shelter, while the hot sun and the moisture provide an abundance of vegetable food in response to a minimum of labour. Hence, no man perishes through lack of energy to secure the prime necessities of life; and there has been no great weeding out of the indolent by severe conditions of life, such as alone can produce an innately energetic race, one that loves activity for its own sake. For the same reason these same peoples also exercise but little foresight, they are naturally improvident; the abundance of nature renders it possible to survive and propagate without any prudent provision for the future.

Contrast with these races the northerly races—in Asia the Japanese, whose energy and industry we all recognise, and the northern Mongols or Tartars, who have so often overrun and conquered with fire and sword the less energetic peoples of the south, or the Goorkhas or Pathans of the highlands of northern India. But more especially contrast with them the English people. M. Boutmy rightly asserts that "the taste for and the habit of effort must be regarded as the most essential attribute, the profound and spontaneous quality, of the race."[1] It is displayed in the English love of sport and adventure and travel; especially in such recreations as mountain climbing, which is pre-eminently an English sport; also throughout our social life, in the intensity of commercial and industrial activity, often carried on ardently by men far removed from any necessity of making money. In our political life, where a vast amount of effort is constantly expended in achieving comparatively small results, we always seem to prefer to achieve any reform by the methods which give scope to and demand the greatest amount of public activity and effort. It is shown also in the immense amount of public service rendered without remuneration, for the mere love of activity and the exercise of power. It is very striking in English colonies in tropical lands, and has been no doubt an important factor in our success in tropical administration and in colonisation.

Boutmy is inclined to attribute to this love of activity, as a secondary effect, the dislike of the mass of Englishmen for generalisations and for theoretical construction; for, he says, these are the results naturally achieved by the reflective mind; whereas the English mind gets no time for reflection, its attention is perpetually drawn off from general principles by its tendency to pursue some immediate practical end. Hence, he says, abstraction is

[1] *Op. cit.*

subordinated to practical ends and does not soar for its own sake. This truth is well illustrated by the fact that all our English philosophers, Bacon, Hobbes, Locke, Mill, Bentham, Spencer, etc., have been practical moralists, and have conducted their investigations always with an eye to concrete applications to the conduct of the State or of private life.

Boutmy regards this love of activity, together with foresight and self-control, as racial qualities engendered by the severity of the climate, working chiefly by way of natural selection. In the prehistoric period more especially, when man had little knowledge of means of protection from climate and hardship such as have been developed by civilised societies, those individuals who were deficient in these qualities must have succumbed to the rigours of the climate, leaving their more energetic fellows to propagate the race.

That there is truth in the view is shown by the fact that the degree to which the love of activity is developed seems to vary roughly with severity of the climate even among the closely allied races of Europe. As we pass northward from the coast of the Mediterranean, we find the quality more and more strongly marked; and it is in accordance with this principle that the dominant power, the leadership in civilisation, has passed gradually northwards in the historic period. Civilisation first developed in the subtropical regions, in which the abundance of nature first gave men leisure to devote themselves to things of the mind, to contemplation and inquiry; while the northern races were still battling as savages against the inclemency of the climate, were still being ruthlessly weeded out by the severities of the physical environment, and so were being adapted to it, that is to say, were being rendered capable of sustained and vigorous effort. But, as the means of subduing nature and of protecting himself against nature have been developed by man, the dominance has passed successively northwards to peoples whose innate energy and love of activity were more highly developed in proportion to the severity of the selection exerted upon many preceding generations.

The severe climate has not been the only cause of this evolution of an energetic active type. No doubt military selection played its part also. The Northern races of Europe, more particularly the Nordic, the fair-haired long-headed race, underwent a prolonged severe process of such military group selection, before branches of it settled in our island; and, among the qualities which must have tended to success and survival in this process, energy and capacity for prolonged and frequent effort, especially bodily effort, must have been one of the chief. Still, even such group selection was probably a secondary result of the direct climatic selection; for it must have been the love of activity and enterprise that led these peoples perpetually to wander, and so to come into conflicts with one another, conflicts in which the more energetic would in the main survive and the less energetic succumb. In part also it must have been determined in the third and the most indirect manner in which physical environment shapes racial qualities—namely, by determining occupations and modes of life, and through these the forms of social organisation, both of which then react upon the racial qualities.

In illustration of this third mode of action of physical conditions, let us take a striking difference of mental quality between the French and the English peoples, and inquire how the difference has arisen; a difference which is recognised by every capable observer who has compared the two peoples and which has been of immense importance in shaping the history of the modern world. I mean the greater sociability of the French and the greater independence of the English, a greater self-reliance and capacity for individual initiative. The difference finds expression in every aspect of the national life of the two peoples. The sociability and sympathetic character of the French, on which they justly pride themselves,[1] is the inverse aspect of their lack of the characteristic English qualities, independence and self-reliance. In political life the difference appears in the centralised organisation of the French nation, every detail of administration being controlled by the central power through a rigidly organised hierarchy of officials, in a way that leaves no scope for initiative and independence in local administrations. Connected with this is the almost universal desire of educated men to become State functionaries, parts of the official machinery of administration, and the consequent excessive growth of this class of persons.

The same quality of the French shows itself in the tendency to prefer the monarchical rule of any man who shows himself capable of ruling, a tendency which constantly besets the republican State with a well-recognised danger. These are not local and temporary manifestations, but have characterised the French nation throughout the whole period of its existence. In the feudal period which preceded its formation, there was considerable local independence; but the feudal system was due to the dominant influence of Frankish chiefs, of the same race as our Saxon forefathers, who overran most of France as a ruling caste but did not contribute any large element to the population, and whose blood therefore has been largely swamped. It appears in the greater violence among the French people of collective mental processes, those of mobs, assemblies, factions, and groups of all kinds. Each

[1] Guizot asserted that, even when new ideas and institutions have originated elsewhere, it has usually been only by their adoption in France that they have been spread through Europe (*A History of Civilisation in Europe.*)

individual is easily carried away by the mass; there are none to withstand the wave of contagion and, by so doing, to break and check its force.

In England on the other hand political activity has always been characterised by extreme jealousy of the central power, and by the tendency to achieve everything possible by local action and voluntary private effort. All reforms are initiated from the periphery, instead of from the centre as in France. Great institutions, the universities, schools, colleges, hospitals, railways, canals, docks, insurance companies, even water supplies and telephones and many other things which, it would seem, should naturally and properly be undertaken by the State, or other official public body, have been generally set on foot and worked by individuals or private associations of individuals. Even vast colonial empires—India, Rhodesia, Canada, Sarawak, Nigeria, North Borneo—have been in the main acquired through the enterprise and efforts of individuals or associations of individuals; the State only intervening when the main work has been accomplished.

In their religion, too, the English are markedly individualistic; our numerous dissenting bodies have mostly dispensed with the centralised official hierarchy which in Roman Catholic countries mediates between God and man, and have insisted upon a direct communion with God; and we have many little churches each of which governs itself in absolute independence of every other. In the family relations the same difference appears very strongly. The French family regards itself, and is regarded by law, as a community which holds its goods in common; each child has his legal claim upon his share, relies upon his family for support in his struggle with the world, and is encouraged by his parents to do so. In the English family, on the other hand, the father is a supreme despot, who disposes of his property as he wills. The children are not encouraged to look for further support, when once they become adult, but are taught that they must go out into the world to seek their fortunes unaided. At an early age, the English boy is usually thrust out of the family into the life of a school in which, by his own efforts, he must find and keep his position among his fellows; and he lives a life which, compared with that of the French boy, is one of freedom and independence. In the distribution of the people on the land we see the same difference of mental qualities revealed. The French peasants are for the most part congregated sociably in villages and small towns; the English farmer builds his homestead apart upon his own domain. And this determines one of the most striking differences in the aspect of the rural districts of both countries. In the towns also the same tendencies are clearly shown; in the separate little homes of the English and in the large houses of the French shared by several families.

It is in the expansion in the world of the two peoples that the effects of this difference are most clearly expressed and assume the greatest importance. The English race has populated a vast proportion of the surface of the world, and rules over one fifth of the total population. Whereas the French people, who have conquered large areas, have never succeeded in permanently colonising any considerable portion of their conquests; and they have failed to maintain their domination in many regions where they have for a time established it. In every extra-European region where they have come into conflict with the English race they have been worsted.

The secret of the difference in the expansion of the two peoples is the difference of innate mental quality that we are considering, enhanced by the differences of custom and of political and family organisation engendered by it.

For, like all other innate tendencies, the two to which we are referring obtain accentuated expression through moulding customs, institutions and social organisation in ways which foster in successive generations just those tendencies of which these institutions are themselves the traditional outcome and expression. Thus, it is the individualistic nature of the political, religious, and family organisation of the English people which, having been engendered by innate independence of character and having in turn accentuated it in each generation, has enabled the people to achieve its marvels of colonisation and tropical administration. We see these tendencies playing a predominant part in the history of every British colony.

The difference was well brought out by Volney, a French observer of the French and English colonists in the early days of the settlement of North America. He wrote "The French colonist deliberates with his wife upon everything that he proposes to do; often the plans fall to the ground through lack of agreement." "To visit one's neighbours, to chat with them, is for the French an habitual need so imperious that on all the frontier of Louisiana and Canada you will not find a single French colonist established beyond sight of his neighbour's home." "On the other hand, the English colonist, slow and taciturn, passes the whole day continuously at work; at breakfast he coldly gives his orders to his wife; . . . and goes forth to labour. . . . If he finds an opportunity to sell his farm at a profit, he does so and goes ten or twenty leagues further into the wilderness to make himself a new home."[1]

It is the French authors themselves who have most insisted upon this mental difference between the French and the English, which seems to be determining a great difference in the destinies of the two peoples; and most of them, while justly valuing the sympathetic and social

[1] I cite these passages after M. Boutmy, *op. cit.*, p. 168.

quality of the French mind, deeply regret its lack of the English independence.[1] There has been no lack of speculation and inquiry as to the origin and causes of this supremely important difference. It is perhaps worth while to glance at some of these attempts.

The most superficial attempt at explanation is to say that the political and social institutions of the French people foster in each individual the social tendencies in question, while the English institutions develop their opposites. It is true; but it obviously is not the explanation of the difference; for that we must go further back, in order to find the origin of these differences of institution.

An explanation a little less superficial is that the domination of the first Napoleon and the strong centralised system of administration established by him accounts for the difference. But the permanence, if not the very possibility, of that system, and the rise to power of Napoleon himself, were but symptoms of this deep-lying tendency of the French mind.

Buckle, recognising the profound difference which we are considering, summed it up in the phrases "the dominance of the protective spirit in France" and of "the spirit of independence in England." He attributed the former partly to the influence of the Roman Catholic Church in France with its centralised authoritative system, partly to the long prevalence of the feudal system of social organisation, under which every man was made to feel his personal dependence upon the despotic power of an independent noble and was accustomed to look to him for all initiative and guidance—was trained to obey a despot, whose absolute jurisdiction and whose title to his lands and rights was unchallenged. The system, he said, culminated in the despotism of Louis XIV, by the subjection

[1] The most frank, perhaps somewhat exaggerated, expression of the difference is *The Superiority of the Anglo-Saxon*, by Ed. Demolins.

of the previously independent nobles to the king, and was revived in a different form, immediately after the great revolution, by Napoleon.

The dominance of the spirit of independence among the English people he would explain also from the character of their political institutions during recent centuries. After recounting the political history of England from the sixteenth to the eighteenth century, and after showing how the people during that period repeatedly succeeded in asserting its liberties against the encroachments of the kings, he wrote—"In England the course of affairs, which I have endeavoured to trace since the sixteenth century, had diffused among the people a knowledge of their own resources and a skill and independence in the use of them, imperfect indeed, but still far superior to that possessed by any other of the great European countries." But he was not wholly satisfied with this explanation; he added—"Besides this, other circumstances, which will be hereafter related, had, as early as the eleventh century, begun to affect our national character and had assisted in imparting to it that sturdy boldness, and at the same time, those habits of foresight, and of cautious reserve, to which the English mind owes its leading peculiarities."

When we turn to this account of the primary cause of English independence,[1] we find that it was, in his view, that the feudal system was established by William the Conqueror in a form different from that obtaining on the continent. The nobles received their lands directly from the king as grants, and all land owners were made to acknowledge their direct obligation to him. The nobles were in consequence too weak to set up their own power against that of the king, and therefore they called the people to their aid in resisting the power of the king; hence, the people early acquired rights and privileges and the

[1] *History of Civilisation in England*, Vol. II, p. 114.

habit of organised resistance to the central authority. "The English aristocracy, being thus forced by their own weakness, to rely on the people, it naturally followed that the people imbibed that tone of independence and that lofty bearing, of which our civil and political institutions are the consequence, rather than the cause. It is to this, and not to any fanciful peculiarity of race, that we owe the sturdy and enterprising spirit for which the inhabitants of this island have long been remarkable."

"The practice of subinfeudation, became in France almost universal." The great lords subgranted parts of their lands to lesser lords, and these again to others, and so on—"thus forming a long chain of dependence, and, as it were, organising submission into a system." In this country, on the other hand, the practice was actively checked. "The result was that by the fourteenth century the liberties of Englishmen were secured," and the spirit of independence had become a part of the national character. That is to say, Buckle maintained that three centuries of a different form of the feudal system sufficed to produce this profound difference between the French and English peoples.

Boutmy also fully recognises the important difference between the innate qualities of the French and English; and he also would explain it as the effect of political institutions since the middle ages, but on lines somewhat different from Buckle's—namely, that England was early ruled by a king invested with great power, and inclined to all the excesses of arbitrary rule. Hence the first need of the people was to fortify themselves against his power. All the law of England carries the imprint of this fear and this defiance. The parliament has been set up against the crown, the judges against parliament, and the jury against the power of the judges; and so, ever since the conquest, individuals have been accustomed to think, and to assert, that their persons, their purse, and their homes are inviolable; and that the State is an enemy whose encroachments must be resisted. This way of thinking has by long usage become instinctive, increasing from generation to generation; until the horror of servitude has become rooted in the Englishman's temperament, and the desire of independence has become a native and primary passion.

Both Buckle and Boutmy agree, then, that the English love of liberty is due to England having been conquered and ruled by a powerful king, and that in France the opposite effect is to be attributed to the same cause—namely, the influence of despotic rulers. Surely this is to reverse cause and effect. If the English people had not already possessed the sturdy spirit of independence when they were conquered by the Norman, his strong centralised rule would only have rendered them still less independent and would have fostered the spirit of protection, as Buckle calls it. If the national characters had been reversed in this respect, how easy it would have been to show that the dependence of the English character was due to the strong rule of a foreign despot, William of Normandy, while the French independence was due to the existence in feudal times of many centres of independent power, the nobles, each capable of resisting the central authority! It was just because this spirit was theirs already that the English people resisted their kings and were able to secure their liberties by setting up institutions congenial to their nature, institutions and customs which have fostered in each individual and each generation the spirit of independence inherited as a racial quality, and which possibly, though by no means certainly, have further intensified the racial peculiarity.

Another cause for the difference of institutions is assigned by Sir Henry Maine. He pointed to the great influence of Roman law upon French institutions; he showed how the French lawyers, brought up in the school of Roman law and holding the Roman Empire as the ideal of a political organisation, threw all their weight upon the side of the monarchy, and in favour of centralised administration. More, perhaps, is due to this influence than to the causes assigned by Buckle and Boutmy; but no one of these alleged causes, nor all of them combined, can be accepted as adequate to explain the origin of the difference of national characters. These authors fail also to make clear how the political institutions can have modified character. Boutmy frankly assumes use-inheritance, which, as I have said, is, in the present state of science, an unwarrantable assumption.

That these qualities of the French and English peoples are innate racial qualities, evolved during the race-making or prehistoric period, is proved not only by the inadequacy of any assignable causes operating during the historic period, but also by the fact that similar qualities are described by the earliest historians as characterising the ancestors, or the principal ancestral stocks, of the two peoples, when they first appear in history. It is proved also by the fact that other branches of the Nordic race have displayed similar qualities, more especially the Dutch and also the Normans, who, though they have long formed part of the French State in the political sense, and have suffered most of the political influences assigned as causes of the spirit of protection, not only displayed the spirit of independence in the highest degree ten centuries ago, but are admitted to be still distinguished from the bulk of the French people by the greater individualism of their character, just as they are still markedly different in physical traits. They offer one of the best examples of fixity of the physical characters of a race. No one can travel in Normandy without being struck by the very marked and distinctive physical type, which, according to all accounts, is that of the Norman who came over to England with the Conqueror; and there is every reason to believe that the mental qualities of the race have been equally fixed and enduring.

Julius Cæsar, Tacitus, and other early historians have described for us the leading qualities of the Gauls on the one hand and of the Teutons on the other. Fouillée in his *Psychology of the French People* has brought together the evidence of these early historians on the point; it shows that the Gauls and the Teutons were distinguished very strongly by the same differences which obtain between the French and English peoples at the present time, especially the difference in respect of independence and initiative, the origin of which we are seeking to explain. The Gauls were eminently sociable people, sympathetic, emotional, demonstrative, vivacious, very given to oratory and discussion, vain and moved by the desire of glory, capable of great gallantry, but not of persevering effort in face of difficulties, easily elated, easily cast down. And, what from our point of view is especially important, they were readily led by the chiefs, to whom they were attached by the bonds of personal loyalty; and they were constantly banding themselves together in large groups, under such leaders as attained popularity by their superior qualities; and, again, they were dominated by the priestly caste, the Druids. The Gauls even had those family institutions which characterise the modern French and which have been held to be the expression of their recently acquired qualities and traditions; namely, the family had the character of a community in which the wife had equal rights with the husband, and the children were regarded also as members of the community having their equal claims upon the family property. And society was bound together by a system of patrons and clients, a system of personal dependence.

On the other hand, the Teutonic people, as described by the same ancient authorities, displayed a decided individualism in virtue of which their social organisation was more rudimentary. The father was supreme in the family, and his power and property descended to his eldest son. They were a more phlegmatic people, but of great energy and persistence. Unlike the Gauls, they were dominated by no priestly caste. The religious rites were conducted by the elder men.

The Gauls were a mixed people of whom the minority, constituting the nobility, were of the tall, fair, long-headed Nordic race; while the majority, the mass of the common people, were of the short, dark, round-headed race. And these, as the numerous observations of the anthropologists show, constitute to-day the bulk of the population, except in Normandy and the extreme north-east of France.

The Teutons or Germans of Cæsar and Tacitus, on the other hand, were of the fair Nordic race; and the Anglo-Saxons who overran Britain, together with the Danes and Normans, who, with the Saxons, formed the principal ancestral stock of the English, were of this same Nordic race, or Northmen, as we may call them.

Now, it might seem useless to attempt to arrive at any conclusions as to the influences that shaped these races in prehistoric times. But an attempt has been made by one of the schools of French sociologists, which, in spite of its speculative character, seems to be worthy of attention. This is the school of "La Science Sociale," founded seventy years ago by Fredericq le Play and more recently led by Ed. Demolins and H. de Tourville. Aided by a number of ardent disciples, they have made a special study of the influence of physical environment in determining occupations and social organisation, and in moulding indirectly through these the mental qualities of peoples. That is their great principle. They rightly, I think, insist upon the relatively small importance of political institutions in moulding a people, regarding them as secondary results of the factors which, determining the private activities of men and women at every moment of their lives from the cradle to the grave, exert a far greater and more intimate influence upon their minds. In two fascinating volumes[1] Demolins has summed up the principal results of this school and attempted to trace the conditions that have determined the differentiation of all the principal races of the earth; and de Tourville has applied the same principles and traced their effects in European history.[2]

It is a curious fact that the work of the Le Play school is almost entirely ignored by the other French sociologists and anthropologists. It is seldom referred to by them, and outside France also it has not received the attention it deserves. Much of it is of the nature of brilliant speculation, and is regarded no doubt as unsound by many more sober minds. Yet, when we attempt to understand the evolution of man in the prehistoric period, brilliant speculation becomes a necessary supplement to the work of measuring skulls and digging up ruins, to which some less ingenious workers confine themselves. And, of all the conclusions of the Le Play school, their account of the origin of the distinctive characters of the Northmen is one of the most striking and satisfactory; while their account of the origins of the Gauls and of their peculiar social organisation and well-marked mental traits is also among their best work.

[1] *Comment la route crée le type social*, Paris, Didot et Cie.

[2] *Histoire de la Formation particulariste*, Paris.

CHAPTER XVII

The Race-Making Period (*continued*)

THE INFLUENCE OF OCCUPATIONS AND OF RACE-CROSSING

IN the foregoing chapter we noticed certain well-marked and generally recognised differences of national character presented by the French and the English peoples—namely, the greater independence of the English, the greater sociability of the French people; and we noted how these differences of national character show themselves throughout the institutions of the two nations, and how they have played a great part in determining the difference of their histories; especially, we saw, how they are of prime importance, when we seek to account for the greater expansion of the English people throughout the world.

We then noticed several attempts that have been made, by Buckle, Boutmy, Maine, and others, to account for these differences as results of differences of political institutions during the last thousand years. We found that all these attempts fail, and that the differences of political institutions, which these authors have regarded as the causes of the differences of national character, are really the expressions of a fundamental racial difference; that, in short, these authors have inverted the true causal relation. I then drew attention to the work of the school of Le Play and especially to its fundamental principle—namely, that, while peoples are in a state of primitive or lowly culture, their geographical or physical environments determine their occupations and, through their occupations, their social organisations, especially their domestic organisation; and that particular modes of occupation and of social organisation of a primitive people, persisting through many generations, mould the innate qualities and form the racial character.

I said that two brilliant workers of the school—namely, Demolins and de Tourville—had applied this principle to account for those differences between the national characters of the French and English peoples which we were considering. I have now to reproduce their account in as condensed a form as possible.

Demolins claims to show that the short dark round-headed people, who formed the bulk of the Gauls and also of the population of modern France, came, in prehistoric times, from the Eurasian steppe region, reaching France by way of the valley of the Danube, a long narrow lowland region confined on the north by the Carpathians and mountains of Bohemia, on the south by the Balkans and Swiss Alps. He supposes that, for long ages, they had lived as pastoral nomads on the steppes. By examining the nomads who still lead the pastoral life on the steppes, he shows the kind of social organisation to which this pastoral life inevitably gives rise and under which they lived; and he traces the effects which such occupation and such social organisation produce on the mental qualities of a people.

The system is the patriarchal system *par excellence*. It is something very different from the Roman system characterised by the *patria potestas*, which the writings of Sir H. Maine have perhaps tended to confuse with the true patriarchal system. The patriarchal system of the pastoral nomads is essentially a communal system, under which all the brothers, sons, and grandsons of the patriarch form, with their families, a community which holds all the property, consisting of flocks and herds, in common; each member having his claim to his share of the produce, each doing his share of the common labour, and each having a voice in the regulation of the affairs of the family. Such a system represses individualism; there is no individual property, there are no individual rights, duties, or responsibilities; no scope for individual initiative; the individual is swallowed up in the community; superior energy or enterprise bring no superior rewards, but rather tend to social disorganisation and to the detriment of the individual who displays them. Further, the work of looking after the herds of cattle is easy and delightful, calling for no sustained exertion; and the herds provide every necessary article of food, clothing, and shelter. Beyond the family group there exists no political organisation; for the group is self-supporting and independent, it has no need of relations with other groups, and each group lives far apart from others, wandering in some ill-defined region of the immense plain.

The peculiarities of this social organisation and of this mode of life are clearly created by the physical environment, by the boundless grassy plains, which enable each family group to maintain a large troupe of cattle, chiefly horses. At the same time, these conditions render necessary the co-operation of all the members of the family in the common work of tending the cattle; while the necessity of continually moving on to fresh pasture prevents the growth of any fixed forms of property and of any more elaborate social organisation.

It is an extremely stable and persistent mode of life and of social organisation. So long as the geographical conditions remain unchanged, it is difficult to see how any change would take place in it, how any progress towards civilisation could begin. And, as a matter of fact, the people who have remained in these regions continue to lead just the same patriarchal, pastoral, nomadic life. Long ages of this mode of life may well put upon a people the stamp of sociability and communism and kill out individualism and individual initiative! Demolins points out in a very interesting way how these effects of the patriarchal system of the pastoral nomads are displayed most clearly still by the population of southern Russia, who, of all the settled European peoples descended from such pastoral nomads, have suffered fewest disturbing influences; how still the individual is subordinated to the community, to the *mir*, by which all private life and industrial activity is directed and which is the owner of the principal property, namely the land; and how, in consequence, the people remain devoid of all individual initiative and enterprise.

The Celts arriving in Gaul retained these qualities and something of the patriarchal organisation, although they were no longer simply pastoral nomads; for, in the course of their migrations, they had been forced to take up agriculture and the rearing of other domestic animals, especially the pig, through lack of sufficient open steppe land. While in this disorganised condition in Gaul, they were overrun by tribes of the Nordic race, who established themselves as a conquering nobility, superimposing upon the rudimentary political organisation of the Celts a loose military organisation of clans; each clan was led by a popular warrior who attached to himself by his personal qualities as large as possible a number of clients or clansmen, acquiring rights over their land and property, in return for the patronage and protection he offered them. These nobles with their blood relatives were the tall fair-haired Gauls described by Cæsar. The Celts lent themselves readily to this system based on personal loyalty and leadership, owing to their lack of independence of character engendered by long ages of the patriarchal communal regime. And the new social organisation fostered and developed still more through many generations the spirit of dependence, the tendency to look for authoritative guidance and control to some recognised centre of power.

In these two circumstances, the long regime of patriarchal communism and the subsequent prevalence for many generations of the clan system, we may see, according to Demolins, the causes of those deep-seated tendencies of the French nation (summed up by Buckle in the phrase, the spirit of protection) which throughout their history have played so large a part in shaping the destinies of the people, and which are still the source of grave anxiety to many patriotic Frenchmen.

It is interesting to note that among the Celtic populations of the British Isles the same features have been clearly displayed. We see among them the clan-system with its dual ownership of the soil, which has been perpetuated in Ireland to the present day and has received more formal and legal recognition from the British government in its recent legislation. We see the strong clannish spirit and relative lack of independence. These qualities are clearly shown by the Celtic Irish, even when they have been compelled by necessity to emigrate to America. There they are not found to be pioneers on the frontiers of civilisation, but rather remain herded together in clannish communities in the cities of the eastern states, where they create such powerful unofficial associations as "Tammany Hall."

Demolins' account of the genesis of the spirit of independence and enterprise of the Anglo-Saxons is still more interesting and seductive. He supposes that their ancestors also came originally in very remote times from the Eurasian steppes; but that is a disputable point and forms no essential part of his argument. They settled in prehistoric times around the coasts of the Baltic and the North Sea, especially in Scandinavia. And the physical peculiarities of this region impressed upon their descendants the qualities which have enabled them to play a leading part in the destruction of the Roman power and in the development of the civilisation of modern Europe, and which have established them in almost every part of the world as a dominant race, increasing in power and numbers at the expense of other peoples.

What, then, are these physical conditions?

Scandinavia is a mass of barren mountains coming down in almost all parts abruptly to the sea. Its coast line is indented by innumerable fiords and bordered by thousands of small islands; and the sea which washes these coasts is warmed by the Gulf Stream. This sea, owing to its warmth and to the existence of a great bank which lies near the surface and runs parallel to the coast line, is extremely rich in fish. Hence, the Nordic tribes who settled in Scandinavia inevitably became a sea-faring folk, spreading slowly along the coasts in small boats, supporting themselves in large part upon the fish which they caught in the sea; for the land is barren, while the sea offers ideal conditions for fishing in small boats. But, unlike the herds of pastoral peoples, sea-fishing does not provide all the necessities of a simple life. It must be combined with agriculture. Hence, the ancient Northmen became a race of hardy seafarers who at the same time practised agriculture.

The character of the land which was available for the necessary but supplementary agriculture was all important. It consisted, as it does still, of small isolated strips of cultivatable soil at the foot of the mountains where they plunge into the sea. On such land it was impossible for the family to retain the form of a patriarchal community. The fertile areas were too small to support such communities, and the individualistic form of family was inevitably evolved. On each small plot of cultivatable land a little farm was formed, a homestead in which lived a family restricted to father, mother, and children. As the children grew up, it was impossible to support them on the one small farm or to divide it among them; one son alone was chosen as the inheritor of the paternal farm; and each of the others had to seek a new piece of land, build a new homestead, and acquire his own boat.

Thus, the family was forced to become the individualistic family; and the home of each such family was necessarily isolated, widely separated from that of every other, owing to the scattered distribution of the little areas of fertile soil. Thus were formed the first homes in the English sense of the word; the home in which the father rules supreme over his own little household, brooking no interference from outside; the home in which the children are brought up to look forward to establishing, each child for himself, similar independent individualistic homes. Such homes have been established by the Northmen in every part of the world in which they have settled; and they are peculiar to them and their descendants.

It is obvious that all the very limited domains of the Scandinavian coasts must have been fully occupied in the way described in a comparatively few generations after the process of settlement began. This seems to have occurred about the fourth or fifth century A.D. Then the younger sons, for whom there was no place at home and for whom there remained no spots suitable for homesteads in their native land, were sent out into the world to seek their fortunes. They banded themselves together to man single boats, or formed fleets of boats; and, leaving their parents and women-folk behind, set out to conquer for themselves new homesteads. Large numbers, sailing to the southern shores of the Baltic and up the Weser and the Elbe, settled on the plains of Saxony; and from this new centre they again spread, as the Anglo-Saxons to England, and as the Franks to Gaul. Others settled directly in northern France and became the Normans. Others, the Varegs, penetrated the plains of Russia and established themselves as princes over the Slav population.

This was a migration such as had never before been seen; bands of armed men, all young or in the prime of life, coming not as mere robbers, but seeking to conquer for themselves and to settle upon whatever land seemed to them most desirable. Everywhere they went they conquered and either exterminated or drove out the indigenous population, as in the south and east of England, or established themselves as an aristocracy, a ruling military caste, as the Franks in the north-east of Gaul. And everywhere they established firmly their individualistic social organisation, especially the isolated homestead of the individualistic family, characterised by the despotic power of the father and by great regard for individual property and for the rights of the individual as against all State institutions and public powers. In hostile countries the homestead became a fortified place, or at least was furnished with a fortified keep or castle; and in those regions, such as Gaul, in which the indigenous population was not exterminated, the feudal system was thus initiated. Everywhere they carried their spirit of independence, enterprise, and initiative.

It was the swarming of the young broods of Northmen in search of new homes that caused the Romans to describe these Northern lands as the womb of peoples, and to regard them with wonder and something of fear.

These qualities and habits continued to be displayed in the highest degree by the Normans after their first settlement in the north of France. The younger sons kept up the good old fashion of going out into the world to seek a fortune or rather a territory, which often was a dukedom or a kingdom. Their most characteristic performance was the conquest of the greater part of Italy. A little before William of Normandy and his companions secured for themselves domains in this country, Norman knights, engaging in enterprises that might well have seemed absolutely foolhardy, had established themselves in Mediterranean lands. Some two thousand Normans, arriving Viking fashion in their small ships, conquered Sicily and the south of Italy and divided these lands among themselves; and for a time they introduced order and a settled mode of life among the peoples of those parts. The leading spirits among them were ten sons of one Norman gentleman, Tancrède de Hauteville, the father of twelve sons of whom two only remained at home, while each of the others carved out for himself a domain in Italy. As Demolins remarks, these families, retaining undiminished their individualistic tendencies and spirit of independence, were veritable factories of men for exportation.

The modern Frenchman, says Demolins, would regard as the height of folly the enterprises of the old Northmen, who, mounted on their frail ships, quitted each spring the coast of Scandinavia, launched out on the wild sea, landed, a mere handful of men, on the coasts of Germany, Britain, or Gaul, and there with their swords carved out domains and made new homesteads. It was thus that the ancestors of Tancred had acquired the manor of Hauteville, and it was thus that his sons conquered Italy and Sicily.

It was in a very similar way that, in a later age, men of the same breed carried to the new world the same individualistic institutions and the same spirit of independence, and in doing so, laid the sure foundations of the immense vigour and prosperity of the American people.

There is one almost more striking illustration of the great and lasting effects upon character and institutions of the mode of life of the Northmen determined by their physical environment. It is furnished by the character and habits of the people who still dwell in the plains between the mouths and lower parts of the Weser and the Elbe, a region which was naturally one of the first to be conquered and occupied by the Northmen. This territory is an infertile sandy plain, and at the time of the coming of the Northmen had but scanty population; hence, instead of becoming the military and ruling caste of a subject people, the Northmen became themselves peasants and farmers. In doing so, they retained all the characteristic features of the individualistic family and have perpetuated them, together with the spirit of enterprise and independence, undiminished to the present day.

In this region each farm is a freehold which has remained in the hands of the same family for long periods, in many cases for hundreds of years. Each farm has its isolated homestead inhabited by the head of the family, his wife and young children, and one or two hired servants. Each homestead is well-nigh completely self-supporting and lives almost independent of the outside world. In spite of the isolation, which might have been expected to engender an extreme conservatism and backwardness of culture, these farmers have continued to exhibit the old Northmen's spirit of enterprise and their power of voluntary combination in the pursuit of individual ends. They were the first in Europe to establish a society for the scientific study of agriculture, and they have thus maintained themselves in the first rank as cultivators of the land, quite without State assistance. In the same way and at an early date they established schools for their children. They have continued to produce large families and have retained the custom of handing over the farm and homestead intact to one son, chosen for his ability to manage it; while all the other sons keep up the old custom of going out into the world to seek their fortunes, in the shape of new homesteads.

Most striking of all, they still do this in the old Norse fashion as nearly as possible. In one district these farmers combined their efforts some sixty years ago and built a ship which, since that time, has sailed every year to South Africa, carrying there the surplus sons in search of new domains for themselves. In that far country their spirit of independence finds satisfaction in establishing new homesteads, new families of the individualistic type, and in perpetuating their traditions of enterprise and self-reliance.

It is because the modern Scandinavians are of the same stock, fashioned for long ages by the same physical environment, that they have continued to emigrate in large numbers to North America, where some of their ancestral race landed centuries before Columbus was born, and where, in the newly opened territories of Canada and the United States, they are generally recognised as being among the best of the settlers.

Demolins does not enter into the question—How did the institutions and mode of life of these or other peoples, determined by physical environment, bring about adaptation of racial qualities to the environment? He seems to assume in all cases use-inheritance. But if, as seems possible or even probable, this is a false assumption, we may still see clearly that, in the case of the Northmen at least, adaptation may well have been effected by selection. The conditions of life of these Northmen were such that in each generation the majority of men could become fathers of families only after carrying through successfully an enterprise in which a bold independence of spirit was the prime condition of success.

Those who were deficient in the spirit of independence must have shrunk from these wild expeditions in search of new homes to be won only by the sword, or must in the main have failed to attain the end; remaining at home, or returning there after failing in the enterprise to which they proved unequal, to finish their days as bachelor uncles at the paternal hearth. This process, carried on for many generations, would lead to the evolution of just those qualities which are characteristic of their descendants in all the many parts of the earth where they now rule. Not only must such social selection have been operative during the period of settlement of Scandinavia; but each great migration to a new area must have sifted out the most independent and enterprising spirits to be the founders and fathers of the new branch of the race.[1] Thus the descendants of the pilgrim fathers were the product of three such processes of severe selection; the migration from Scandinavia to Northern Germany; that from Germany to England; and that from England to America. No wonder that they proved themselves well able to cope with the hardships and dangers of a new continent inhabited by savages only less fiercely tempered than their own stock by many generations of warfare! When we thus find the same institutions and the same mental traits characterising, from the dawn of history to the present time, all the widely separated branches of one racial stock and of this stock alone, we realise how powerful over the destiny of nations is the influence of racial character formed in the long prehistoric ages; we see how futile it is to attempt to explain the mental traits of a people by the history of their political institutions during a few recent centuries; we understand that these institutions are

[1] The reality of selective effect of migration is shown by the stature of American immigrants; those from Scotland are said to be two inches taller than the average Scotchman; and De Lapouge shows (*Les Sélections Sociales*, Paris, 1896, p. 367) that a superiority of stature almost as marked, may be inferred for the French and German immigrants of America from the statistics of the armies of the Civil War.

the effects, not the causes, of those mental qualities and that, even among the peoples who have attained the highest degree of civilisation, racial qualities remain of supreme importance.

The Crossing of Races

Before passing on to the consideration of evolutionary changes during the historic period, a few words must be said about the crossing and blending of races. Such blending has been, no doubt, one of the principal causes of the great variety of human types at present existing on the earth. It has been going on for long ages in almost all regions; but especially in Europe and Africa. All existing stocks (with few exceptions) are the products of race-blending. No one of the existing European peoples is of unmixed stock; every one is the product of successive mixtures and blendings of allied stocks; and the mixing and blending still goes on; while in America (both north and south) the greatest experiments in race-blending that the world has yet seen are taking place before our eyes.

Authors differ widely as to the results of the crossing of human races and subraces. Some assert that the effect of crossing of races is always bad, that the crossbred progeny is always inferior to the parent stocks. They make no allowance for unfavourable conditions, especially the lack of the strong moral traditions of old organised societies. Others maintain the opposite opinion. Both opinions are probably correct in a certain sense. I think the facts enable us to make with some confidence the following generalisation. The crossing of the most widely different stocks, stocks belonging to any two of the four main races of man, produces an inferior race; but the crossing of stocks belonging to the same principal race, and especially the crossing of closely allied stocks, generally produces a blended subrace superior to the mean of the two parental stocks, or at least not inferior.

This generalisation cannot yet be based on exact and firmly established data, unfortunately; but it is in harmony with old established popular beliefs, and with what we know of the crossing of animal breeds; and it is borne out by a general inspection of many examples. For instance, the blending of the white, Negro, and American stocks, which has been going on in South America for some centuries, seem to have resulted in a subrace which up to the present time is inferior to the parent races; or at any rate to the white race. So the mulattoes of North America and the West Indies, although superior in some respects to the pure Negroes, seem deficient in vitality and fertility, and the race does not maintain itself. The Eurasians of India are commonly said to be a comparatively feeble people. The blend of the Caucasian with the yellow race is also generally of a poor type. Examples abound in Java of people of mixed Javanese and Dutch blood; and they are for the most part feeble specimens of humanity. It is generally recognised that a recently blended stock may produce a few individuals of exceptional vigour and capacity and physical beauty. But setting these aside, the blended stock seems to be inferior in two respects: (1) a general lack of vigour, which expresses itself in lack of power of resistance to many diseases and in relative infertility; so that the blended stock can hardly maintain its numbers; (2) a lack of harmony of qualities, both mental and physical. It may be that such lack of harmony is the ground of the relative infertility of blended stocks. It expresses itself in the inharmonious combination of physical features, characteristic of the mongrel. The Negro race has a beauty of its own, which is spoilt by blending.

As regards mental constitution, although we cannot directly observe and measure these disharmonies of composition, there seems good reason to believe that they exist. The soul of the crossbred is, it would seem, apt to be the scene of perpetual conflict of inharmonious tendencies. This has been the theme of many stories, and, though no doubt many of them are overdrawn, there is no reason to doubt that they in the main depict actual experience or are founded on close observation.

It is on the moral, rather than the intellectual, side of the mind that the disharmony seems to make itself felt most strongly; and the moral detachment of the crossbred from the moral traditions of both the parent stocks is possibly due in part to a certain lack of innate compatibility with those traditions, as well as to social ostracism; the crossbred can assimilate neither tradition so easily and completely as the pure-bred stocks.

It is possible, though this is a still more speculative view, that the same is true of the intellectual constitution of the mind.

The superiority of subraces formed by the blending ot allied stocks seems to fall principally under two heads: (1) a general vigour of constitution; (2) a greater variety and variability of innate mental qualities. The greater variability of qualities of a subrace renders that race more adaptable to changing conditions; for racial adaptability depends upon the occurrence of abundant spontaneous variations. A large variety of innate qualities renders a race capable of progressing rapidly in civilisation; it renders it more capable both of producing novel ideas and of appreciating and assimilating the ideas, discoveries, and institutions of other peoples; and such imitative assimilation from one people to another has been a main condition of the progress of culture.

It is, of course, well recognised that the great centres of development of culture have been the places where different peoples have come most freely into contact, notably the centre of the old world where Asia, Africa, and Europe meet together. This was the area in which the three great races of Europe came first into contact and mingled freely. Some authors attribute the fertilising influence upon culture wholly to the blending or contact of cultures; but there is good reason to believe that it is largely due also to race-blending.

We might compare in this respect the three great culture areas of the old world—Europe, India and China. The Chinese afford an instance of one relatively pure race occupying a very large area. In spite of its early start and great mental capacities, its culture has stagnated. The stock was perhaps too pure. India on the other hand seems to owe its peculiar history largely to the fact that its population in almost all parts has been made up from very widely different races—white, yellow and black; the heterogeneity has been too great for stability and continued progress. In Europe different branches and subbranches of the white race, that is of stocks not too widely different in constitution, have undergone repeated crossing and recrossing.

It is worth while to point out that, if our generalisation is valid, it follows that race-blending has been an important factor in the progress of civilisation. And the generalisation has also an important bearing upon one of the most urgent problems confronting the statesmen of the world at the present time, and not only the statesmen but all the citizens of the civilised states, especially the citizens of the British Empire and of America. For it justifies abundantly the refusal of the white inhabitants of various countries to admit immigrants of the yellow or Negro race to settle among them; and it justifies, and more than justifies, the objection to intermarriage with those other races which Englishmen have upheld wherever they have settled, and which most other peoples have not upheld.[1]

In all the currents of heated discussion as to the rights and wrongs of the treatment of other races, this question of the kind of subrace which will result from intermarriage is generally left in the background; whereas its importance is far greater than that of all other considerations taken together. Some, like Sir S. Olivier,[2] are content to approve race-blending on the ground that it improves the inferior race. But the racial qualities of the leading peoples of the world are too precious to be squandered in the process of improving in some uncertain degree the quality of the overwhelming mass of humanity of inferior stocks; the process would probably result in the total destruction of all that humanity has striven and suffered for in its nobler efforts.

It is an interesting question—When two races or subraces are crossed, do they ever produce a homogeneous and true subrace, exhibiting a true and stable blend of the qualities of the parental stocks? Or does the blend always remain imperfect, with many individuals in whom the qualities of one or other of the parental stocks predominate? The answer seems to be that a stable subrace may be formed in this way, though usually not until free intermarriage has gone on for many generations. According to the most recent doctrine of heredity, the Mendelian, every human being is a mosaic or patchwork of unit quali-

[1] A. Reibmayer (*Inzucht u. Vermischung beim Menschen*, Leipzig, 1897) insists upon the importance of isolation and consequent inbreeding for the formation of superior strains and subraces. He points out that the geographical barriers of Europe have favoured in this way the production of distinctive national types. Like Stewart Chamberlain, Flinders Petrie, and others, he regards the dark ages of Europe as a period of chaos directly due to the overcoming of these geographical barriers and the consequent prevalence of crossbreeding on a large scale.

[2] *White Capital and coloured Labour.*

ties, organs, or capacities, each of which is inherited wholly from one of the parents and not at all from the other. If this view is well founded, it follows that there can be no true blending of these unit qualities. But still the mosaic may be so finely grained and the unit qualities derived from the two parents so closely interwoven, that each individual may present an intimate mixture of the parental qualities, may represent for all practical purposes a blending of the two stocks.

CHAPTER XVIII

Racial Changes During the Historic Period

WE have found reason to believe that national character, as expressed in the collective mental life of any people, is only to be understood and explained when we take into account the native or racial mental qualities of the people; and we have seen reason to think that these racial qualities were in the main formed in the prehistoric or race-making period; we have noted some of the principal attempts to throw light on the prehistoric moulding of races. But these racial qualities, although very persistent, are not unalterable. We must, therefore, consider whether, and in what ways, the racial mental qualities of a people may have been changed during the nation-making or historic period. What are the factors which determine such changes? What is their influence on the destiny of nations?

The most diverse opinions are still held in regard to the question of the extent and nature of changes of innate mental qualities of peoples during the historic period, the period during which a people, or a branch of a people, attains political unity and becomes a nation.

There is no doubt that the moulding power of physical environment tends to become greatly diminished during this more settled period of the life of a people, and that, in so far as changes take place, they are determined principally by racial substitutions and by social selections within a people, rather than by the mere struggle for survival of individuals or of family groups against the inclemency of nature or against other individuals and groups.

The former of these two modes of change, substitution, has undoubtedly been effected on a large scale, producing in certain instances radical changes in the racial quality of the populations of some countries; that is to say, there has been more or less gradual substitution of one race for another, while the nation as a geographical and political entity, with its language and much of its laws, institutions, and customs, lives on without complete breach of continuity, and the people, although by blood radically changed, continues to regard itself as the same people, accepting as its own the traditions of those predecessors whom they mistakenly regard as their ancestors.

Perhaps the most striking and complete change of this sort in European history was the change of racial character of the Greek people. It is now pretty well established that the Greek population of the classical age was an incomplete blend of two of the three great European stocks, namely *Homo Europæus*, the Northern, fair, long-headed type of tall stature, and of *H. Mediterraneus*, the short dark long-headed type of the Mediterranean coast lands. The Pelasgians, who, as we now know, had achieved a civilisation of a type that was widely spread through southern Europe as long as three thousand years or more before our era, seem to have been of this Mediterranean race.

Rather more than a thousand years before our era, the Pelasgian population of Greece and the neighbouring regions began to be overrun and conquered by tribes of the fair Northern race which came in successive invasions, the Thracians, the Hellenes, the Achæans, the Ionians, and later the Dorians. Just as, at a later period, men of the

Northern race established themselves as a military aristocracy over the Celtic peoples of western Europe, so these invading tribes established themselves as a military aristocracy over the populations of Mediterranean race; and, as in the former case, so here, they intermarried largely with the people they conquered and formed an imperfectly blended population, in the upper social strata of which the fair type was predominant, in the lower strata the dark type.

From this happy blending of two races was formed the people which, under the favourable geographical and social conditions of that time and place, evolved the civilisation that culminated after six hundred years in the Athenian culture of the time of Pericles. And then, after a very short time, the whole of that splendid civilisation faded away, and the Greek people sank to a position of slight importance from which it has never again risen. After having displayed in several departments of the intellectual life a power and originality such as have never been approached by any other people, they became a people of very mediocre capacities, devoid of power of origination and purely imitative.

That this profound change in the mental qualities of the population of Greece was due to substitution of one racial stock by an inferior one is beyond question. That a great change of racial type was effected is sufficiently proved by the comparison of the physical type of the modern with that of the ancient Greeks. The modern are predominantly dark and round-headed; the ancient were distinctly long-headed, as shown by a sufficient number of skull measurements; and they were, as regards the dominant class at least, predominantly fair in colour. It has been supposed that the many references to the fair hair and complexion of heroes and gods were due to fair persons being very rare and hence an object of special admiration;

but there is no ground for this. The way in which this racial substitution took place is also pretty clear; and the rapid, almost sudden, decline of the intellectual productivity of the Greek people coincided in time with the racial change.

The first and most important factor in the extermination of the best blood of ancient Greece was military selection. Military group selection in the prehistoric period had, no doubt, played a great part in bringing about the evolution of the superior mental qualities of the European peoples, especially of the fair Northern race. So long as the peoples consisted of more or less wandering tribes of pure race, which waged a war of extermination upon one another, the peoples and tribes of superior mental and moral endowments must in the main have survived, while those of inferior endowments went under. But, so soon as the Nordic tribes became settled as aristocracies ruling over the Pelasgian populations, the effects of military selection tended to be reversed; instead of making for racial improvement, they made for deterioration. That racial deterioration occurs under these conditions seems to be an almost general law; it has been exemplified among many different peoples. The many small Greek states were almost perpetually at war with one another; and the result of the warfare was not so much the wholesale extermination of the people of any one state, as the killing off in large numbers of the younger men of the ruling caste, the free citizens of whom the armies were almost entirely composed.

The wars between Sparta and Athens were the most destructive and tragic of all in this respect. We know that the numbers of Spartans of the aristocratic class, never very large, became fewer and fewer, in spite of efforts made to keep up the number by admitting to citizenship persons not of pure Spartan blood; and that

Sparta was eventually destroyed simply for lack of men, men of the ruling class.[1]

In Athens and other states the depleting agencies were more numerous. Frequent wars played the same part as in Sparta; and the number of free men was further diminished by the repeated founding of colonies, in which a relatively small number of persons of Greek blood became swallowed up in a large population of mixed and inferior origin. In some states, in Athens especially, the political conditions worked powerfully in the same direction. Prominent citizens were perpetually exiled or condemned to death, sometimes in considerable batches. It is said that at certain times two thirds of the citizens of certain states were living in exile; and the exiles, going to the colonies or other foreign lands, were for the most part lost to the Greek people.

Then, with the blooming period of Greek intellect, came the loss of the ancient religious beliefs, beliefs which had strengthened the family and made each man anxious to have many sons that the rites might be duly performed for the repose of his shade. Coinciding with this was the great increase of luxury which made large families too expensive, save for the most wealthy; while at the same time the abundance of slave labour kept down the rate of remuneration of all handicrafts, and so condemned the class of free Greek artisans to a state bordering on poverty. Hence, the free citizens of pure blood, already largely reduced in numbers, ceased to multiply; and the number of citizens was sustained only by the admission to citizenship of foreigners, freed slaves, and various elements of different and inferior racial origins.

Hence, at the time that the battle of Chersonese was

[1] Aristotle says "want of men was the ruin of Sparta." Fathers of three sons were exempted from military service, and of four sons from all State burdens.

fought and the Macedonians attained the supremacy, the Greek citizens were no longer the same racial aristocracy which had produced the finest flowers of Greek culture. But the work of substitution was still only partially accomplished. In the time of the Roman domination of Greece, the remnants of the true Greek aristocracy were removed by the slave trade. Tens of thousands of Greeks of all classes were brought together to the slave markets; while those men of talent who escaped that fate emigrated to Rome to seek their fortunes by teaching the Greek language and art and philosophy. Later still came the Goths, who sacked the towns and destroyed or drove out the inhabitants. Then followed successive invasions of Slavs from the north; and lastly, the domination of the Turk well-nigh completed the extinction of the old aristocracy.

The modern Greek people is descended largely from Slav invaders and largely from the numerous and prolific slave population of the great age of Greece, but hardly at all from the men who made the greatness of that age.

Though the change and deterioration of the racial mental qualities of the Greek people by racial substitution is the most striking example in history, it is by no means the only one.[1]

The substitution in that case was largely by elements drawn from other regions and peoples. But a similar substitution and consequent change of innate mental qualities may go on slowly within any people which has been formed, as have almost all the present European nations, by an incomplete blending of two or more racial stocks; it may be effected by internal selection without

[1] Several writers have pointed out the importance of these facts and at least one professional historian has insisted strongly upon them, namely O. Seeck in his *Geschichte des Untergangs der antiken Welt*, Berlin, 1910, vol. i.

any introduction of new elements from any other region. Before considering an example of the process, let us note certain facts which show that there may well have taken place, throughout the historic period, changes of the composition of peoples by internal substitution or changes of the mental constitution by internal selections—that is to say, by the more rapid multiplication of certain mental types and the relative infertility of other types. Consider first the striking fact that the populations of the various European countries seem for the most part to have remained almost stationary as regards numbers, or even in some cases to have diminished greatly in numbers, throughout the period between the Roman domination and the later part of the eighteenth century. The population of Spain is said to have declined from forty millions under the Roman rule to only six millions in the year 1700 A.D. The population of Great Britain is said to have increased from five millions to six millions only during the seventeenth century; and it is certain that in the main it had increased at an even slower rate, or not at all, in the preceding centuries since the Norman Conquest; whereas in the nineteenth century it increased from thirteen millions to nearly forty millions; that is to say, it trebled itself in the century; and even that rate of increase is considerably less than the possible maximal rate.

The same is roughly true for most of the European countries; their populations, throughout great stretches of the historic period, remained stationary or increased only very slowly. Now when, during any period, a population does not multiply at the maximal physiological rate, changes of its character may well be taking place; for, in proportion as the rate of increase falls below the maximal, there is a lack of fertility in the population or in some part of it; if this relative infertility affects equally all parts and classes of the population, it will produce no change of its composition; but if it is selective, if for any reason it affects one class, or persons of some one kind of temperament or mental type, more than others, then this class or this temperament or this form of ability tends rapidly to diminish and to disappear from among that people.

The causes of the relative infertility may be divided into two classes: (1) those which operate by killing persons before they have completed their middle life; (2) those which restrict fertility without killing. Both may be selective in their action. The former kind is alone operative in determining evolution in the animal world and probably also among the less civilised peoples; but, as civilisation advances, the causes of infertility of the other kind increase constantly in effectiveness, while the former operate with less and less intensity. It is through the causes which diminish fertility merely, rather than exterminate individuals, that changes of racial quality of nations are now being, and in the future will be, principally determined. Selection of this kind is usually distinguished from the various modes of natural selection which work by extermination, by the name "reproductive selection." Briefly, natural selection operates by means of selective death rate, reproductive selection by means of selective birth rate.

No doubt, disease, especially in the form of plagues and epidemics, was one of the principal causes of the slowness of increase of population throughout the Middle Ages. And this was probably non-selective as regards mental qualities, although it was strongly selective as regards power of resistance to disease, and has left the European peoples more resistant to most diseases than any other peoples, save perhaps the Chinese.[1]

[1] On this topic cp. Dr. Archdall Reid's *The Present Evolution of Man* and his *Principles of Heredity*, in which books the effects of selection by disease and by alcohol are vividly set out.

But many other causes of selection were at work. Disease presumably has not affected mental qualities by selection; although by direct action and mental discouragement it may have tended to the decay of civilisations; it has been argued, for example, that malaria played a great part in the decay of classical antiquity, that it was introduced some centuries B.C. and enfeebled the population of Greece and Italy.

More interesting, from our point of view, of the influences affecting the mental constitution of populations, is the effect of alcohol. Dr. Archdall Reid has argued very forcibly that resistance to the attraction of alcohol is a mental peculiarity which a race acquires only through long exposure to the influence of abundant alcohol; that populations are resistant just in proportion to their past exposure to it—as is true in the main of epidemic and endemic diseases—and that in both cases this is due to selection.[1]

Much careful painstaking work by continental anthropologists seems to have proved that a change of racial composition through internal selection has been and still is going on in both the German and French people. The facts have been worked out by O. Ammon,[2] Hensen, and De Lapouge.[3] They show, chiefly by means of the comparison of the forms of large numbers of skulls, that throughout the historic period the French and German peoples have been becoming more and more round-headed, that the type of *Homo Alpinus*, the short dark round-headed race, has been gaining upon the type of *Homo Europæus*.

[1] *Op. cit.*

[2] O. Ammon, *Gesellschaftsordnung und ihre natürlichen Grundlagen*, Jena, 1900.

[3] De Lapouge, *Les Sélections Sociales;* cf. also W. Alexis, *Abhandlungen zur Theorie der Bevölkerungs- und Moralstatistik*, Jena, 1903, and W. Schallmayer's *Vererbung und Auslese in ihrer soziologischen und politischen Bedeutung*, Jena, 1910.

We have seen that the latter stock of the fair northern type constituted the upper class among the Gauls of Cæsar's time; and the invasions of Franks and Normans must have added considerably to their numbers; yet, in spite of that, the mental and physical characters of this race are said by these authors to be now very much rarer than formerly, owing to the internal selections which have favoured the Alpine type. These took the following forms. In the first place, in the early Middle Ages, it is said, the Nordic type, being a military aristocracy, suffered, as in ancient Greece, proportionally far greater losses in warfare than the Alpine type. Secondly, the severe persecutions of Protestants in France drove into exile, besides killing many others, large numbers who were for the most part of the fair race, because, as we have seen, this race does not easily remain content within the Roman Church. It is said, for example, that, after the revocation of the Edict of Nantes, so large a number of Protestants passed into Prussia that the rise of Prussia as a powerful State was the immediate consequence, with of course an equivalent loss to France. De Lapouge considers this the greatest blow that France has suffered in the historic period. Normandy alone, it is reported, sent 200,000 Protestants to Prussia. But the most important and curious factor has been, according to De Lapouge, what he calls the selection by towns. He shows, by comparison of masses of anthropological observations, that the Nordic type has been predominantly attracted to the towns (which fact he attributes to their more restless enterprising character) while the dark type has been more content to lead the quiet agricultural life.[1] He points out that the town-life stimulates to a new struggle for adaptation, from which differentiation of classes results; the long-heads maintain their numbers better and rise in the social scale.

[1] Ammon's Law.

Further, he shows that town-life makes against fertility, owing to a number of psychological influences—the stimulation of ambition and of the intellect, the luxurious habits, the weakening of family life, the break with the past and its family traditions, the uncertainty of the future, the weakening of religious sanctions; and he gives reason to believe that in this way the towns have been, through many generations, weeding out the elements of the fair race and determining an ever increasing predominance of *Homo Alpinus*.

In order to understand the importance of these internal selections, it is necessary to realise that their effects are cumulative in a high degree, when the same influences continue to work through many generations. Thus, if within any people there are two equally numerous classes of persons of different mental constitution, *A* and *B*, and if these constitutions determine that the one group *A* has a net birth rate of three children per pair of adults, while the other *B* has a birth rate of four per pair of parents; then, in the third generation after one century, the numbers of the two classes, other things being the same for both, will be as ten to sixteen. After two centuries the one class will be more than twice as numerous as the other; and after three centuries the numbers of the class *A* will constitute about fifteen per cent. only of the whole population. Late marriage is also very important. Suppose that of two classes, *A* marries at 35, and *B* at 25 years, and that each produces four children per marriage; then (other things being the same) after three and a half centuries *B* becomes four times as numerous as *A*. These two factors generally work together.

But, apart from the change of racial composition of a heterogeneous nation by internal selection of this sort, changes of the constitution of even a racially homogeneous people may be produced through selection affecting persons of particular mental tendencies. One of the most striking instances of this is the elimination of the religious tendencies from the constitution of a people by negative selection through the action of the Roman Church.[1] For many centuries the Roman Church has attracted to her service very large numbers of those who were by nature most religiously minded, and it has imposed celibacy upon them, it has forbidden them to transmit their natural piety to descendants. In Protestant countries this process of negative selection of the religious tendencies was continued for a much briefer period than in the Catholic countries. It is maintained with much plausibility that we may see the result in the fact that sincere and natural piety is far commoner in the Protestant countries than in the Roman Catholic; that in the two countries, Italy and Spain, in which the influence of the Roman Church has been greater than in any others, the people are now the least religiously minded of any in Europe; that with them religion has become purely formal and external, that the mass of the people, though outwardly conforming, is absolutely irreligious; that in fact this form of religion tends to exterminate itself in the long run by insisting upon that form of reproductive selection.[2]

Another striking instance of the incidence of negative selection upon certain mental qualities of a people is afforded by the history of Spain. In the sixteenth century Spain attained to a supreme position of power and grandeur among the nations of the world, such as has been rivalled by Rome alone in all history; and then very rapidly her power decayed, and ever since she has remained

[1] Pointed out by Francis Galton and Fouillée.

[2] On the other hand it tends (partly no doubt by deliberate design) to spread itself by insisting upon the duty of procreation. This effect is said to be very considerable in French Canada and only to be partially counteracted by a very high rate of infantile mortality.

one of the most backward of European peoples, contributing little to European culture, to science, art or philosophy, incapable of developing without the aid of foreigners her rich industrial resources, impotent in war, entirely devoid of enterprise and originality. To what is this great change due?

It is not due to any adverse change of climate, to devastation by war or plague or famine, nor is it due to any change in geographical or economic relations. Spain remains more happily situated as a centre of commerce than any other country of the world. The mass of the people remains vigorous, proud, and virile. It is the intellect of the nation alone which has decayed, or rather it is the intellectual life of the nation that has become utterly stagnant.

Buckle drew a vivid picture of the stagnation of the Spanish intellect and sought the explanation of it in the great power wielded by the Roman Catholic Church, which, he said, had successfully fostered the spirit of protection and superstition, had discouraged every effort of the intellect, and utterly repressed the spirit of inquiry, to the free activity of which all progress of civilisation was, in his opinion, due. Here, again, modern science shows that Buckle was led into error by his ignorance of the importance of the biological factors, the racial qualities and the changes produced in them by selection.

Galton and, still more fully, Fouillée have shown that the stagnation of the intellect of the Spanish people and the consequent decay of the power and glory of Spain have been chiefly due to the fact that the people of Spain ceased to produce those men of exceptional mental endowments, of intellectual energy and enterprise and independence of character, on whom primarily depend the power and prosperity of any nation and who are the most essential factors in the progress of the civilisation of any people, who in short are essential for the growth and endurance of national mind and character. And this was because during some centuries intellectual power, enterprise, and energy were steadily weeded out by a rigorous process of negative selection. In the first place, the Church, having attained enormous power, became in two ways a tremendous agency of negative selection. First, she made celibate priests of a very large proportion of all those whose natural bent was towards the things of the mind, multiplying monastic orders excessively. Secondly, by means of the Inquisition she destroyed with fire and sword or drove into exile through many generations all those who would not conform to her narrow creed, who combined intellectual power with independence and originality of spirit and a firm will. In addition she drove out all the Jews and all of Moorish origin.

The second mode of negative selection, namely persecution exerted by the Church, was no doubt the more important, but the former also must have had a great effect. We are helped to realise the probable magnitude of the effect by reflection on facts set out in an article by Bishop Welldon.[1] He shows the great part played in English civilisation since the Reformation by the sons of the English Clergy; including as they did a number of men of the highest achievements in all departments of our national life. If all those sons of clergy who have shown exceptional abilities, and all their descendants, had by the rule of celibacy been prevented from coming into existence, how disastrous would that have been for the English people, how much less successful and vigorous would the nation have become!

A second powerful agency of negative selection was the immense colonial empire which Spain so rapidly acquired, especially her American conquests. The people

[1] *The Nineteenth Century* for April, 1906.

were seized with the desire to enrich themselves with the gold of the New World, and were fascinated by the idea of imitating the romantic adventures of Cortes and Pizarro. Great numbers of the bolder and most capable spirits set out for the New World, and there either lost their lives or remained to mix their blood with that of the native Indians or the imported Negroes. In either case their stock was lost to the mother country.

The third and culminating cause was the career of military aggression pursued by Charles V; this completed the extermination of the aristocracy of ability and finally plunged Spain into an intellectual torpor which has persisted ever since and from which she can be raised up only by a succession of men of first-rate intellect and character: men such as she seems incapable of producing, because her people has thus been drained of all its most valuable elements, because her eugenic stocks have been exterminated.

The fall of Spain illustrates not only the operation of internal social selection affecting certain mental qualities; it illustrates also once more, even more clearly than the fall of Greece, the fact that the civilisation of a people and its power and position in the world depend altogether upon its intellectual aristocracy, and that the fall of a people from a high place necessarily follows the failure to continue to produce such an aristocracy.

In the civilised nations of the modern world, the most important kind of selection at work at the present time is what is distinguished as "economic selection" working in conjunction with the formation of the social classes. It has no doubt operated at various times among other civilised peoples, but never so strongly and universally as at present.

All the leading civilised nations have passed, in the eighteenth and earlier part of the nineteenth centuries, through a period in which the discoveries of science have enormously increased the productive powers of man and man's control over, and power of resistance to, the forces of nature. The result has been that everywhere civilised populations have multiplied at a great rate, in a way that has never before occurred. But now this period seems to have definitely come to an end, and to have been succeeded by a new period characterised by three features which threaten to exert a most deleterious effect upon the innate mental qualities of peoples.

(1) The world is becoming filled up; the untouched wealth of enormous territories no longer lies open to the grasp of the bold and enterprising. The coloured races are entering into the economic competition, in the way foreshadowed by the late C. H. Pearson.[1] The high organisation of every form of economic activity renders the competition for wealth everywhere extremely severe. And at the same time men have come to regard as necessities of life what, but a few generations ago, were the luxuries of the wealthy or unknown even to them; that is to say, the standard of comfort has risen greatly. The combined result of these changes is the increased difficulty of maintaining a family in the upper strata of society.

(2) There has been a great development of humanitarian sentiment, one result of which has been the breaking down of class-barriers and the perfecting of the social ladder; at the same time it has produced such changes of our laws and institutions as tend in an ever increasing degree to lighten the economic burdens of the poor and to consummate by social organisation the abolition of natural selection; that is to say, these changes are putting a stop to the repression by natural laws of the multiplication of the less fit, those least well endowed mentally and physically.

[1] In *National Life and Character*, a pessimistic though intellectually stimulating book.

The recent great decline of infant mortality is one evidence of this.

(3) The influence of religion and custom has weakened, and men are more disposed to adopt the naturalistic point of view, to believe that this life is not a mere preparation for an infinitely longer life elsewhere, but that it is all they can certainly reckon upon and, therefore, is to be made the most of; while at the same time they are oppressed by the severity of the economic competition and by a sense of the lack of any ultimate purpose, end, or sanction of human effort.

The combined result of these three changes is a strong tendency to reverse the operation by which nature has secured the evolution of higher types of mind—namely, by breeding in the main from the higher types in each generation. We see a tendency for the population to be renewed in each generation preponderantly from the mentally inferior elements, those whose outlook hardly extends beyond the immediate future and who have not learnt to demand for themselves and their children favoured positions in the great game of life. The effects of these three changes operate in the following manner. The rate of reproduction, the birth rate, of nearly all civilised countries is falling rapidly (although the death rate also falls). This diminution of rate of reproduction is due to increase of celibacy, abstention from marriage, to increase of late marriage, and to voluntary restriction of the number of the family in marriage.

Now, it is shown statistically that this falling off of fertility chiefly affects the classes above the average of ability, the upper and middle classes and also the superior part of the artisan classes.[1] These classes have been formed and are maintained by the operation of social and

[1] *Provident Societies*, by Sidney Webb, and *The London Population*, by D. Heron.

economic competition; they have long been, and are still, perpetually recruited in each generation from the lower strata, by the rise into them of the abler members of the lower strata. Hence, economic selection, under our present social system, seems to be working strongly for the mental deterioration of the most highly civilised peoples; the social ladder, becoming more nearly perfect, perpetually drains the mass of a people of its best members, enabling them to rise to the upper strata where they tend to become infertile.[1] Galton and Prof. Karl Pearson have insisted most strongly upon these tendencies. But they have not escaped the notice of continental authors. M. Jacobi[2] has written a large volume packed with historical illustrations to prove inductively the law that aristocracies always die out, or are only maintained by constant recruiting from below, or in other words that aristocracies tend to become infertile. And the modern tendency which we have just now considered under the head of economic selection may be regarded as falling under the head of this law, a case of the extension of the law to democratic communities and the natural aristocracies of ability which are generated in them.

We may perhaps state the principal causes of this tendency in general terms as follows: the acquirement by any class of leisure, culture, and the habit of reflection (the malady of thought) partially emancipates that class

[1] It must be recognised also that in Great Britain emigration has, during the last three centuries, tended in all probability in the same direction as the various forms of social selection—namely, to the deterioration of the home population; for in all ages it is the bold and enterprising persons who seek new homes in far countries, leaving the weakly, the timid, the dull, and the defective behind in the mother country. Even the convicts that we exported at one time to our colonies were probably persons of more than average capacity, though some of them may have been innately defective in moral disposition.

[2] *Études sur la sélection chez l'homme.* Paris, 1904.

from the empire of instinct, custom, and the religious sanctions of morality; and these are the great conservative agencies under the influence of which men not so emancipated continue to multiply according to the law of nature. These instincts, customs, and religious sanctions of morality, which lead men to multiply freely, have been acquired for the good of the race or of the society considered as an organism whose life is of indefinitely long duration; and in some respects they are opposed to the pleasure and welfare of the individual life. The habit of reason and reflection tends to lead men to act for their own immediate welfare, rather than for the future welfare of the race or of society, and to refuse to make those sacrifices of ease and to undertake those responsibilities and efforts which the care of a family imposes and which alone can secure the welfare of the future generations. It is in respect to these duties that the great antagonism between religion and reason appears in its most significant aspect.

The tendency for the upper classes to die out and to be replaced constantly from the lower social strata by the aid of the social ladder is no doubt stronger now than in foregoing ages. But it has always been operative; and this is widely recognised; while the comfortable inference has often been drawn that the process is not only inevitable but actually beneficial and desirable. It is said that the upper classes inevitably become effete, and that the lower constitute an inexhaustible reservoir of mental and moral excellences, from which the former can be indefinitely renewed; and thus the population is always rising in the social scale, a state of affairs which makes for social happiness.

But, if we take a longer view, the prospect is not so comforting; it seems only too probable that this constant dying away of society at the top and the renewal of the upper strata from the lower, by the agency of the social ladder, must sooner or later result in a serious deterioration of the lower strata, at least in draining it of its best stocks. There is also a return or downward current of less strength which returns to the lower strata the failures, the incompetents, and the degenerates of the upper. And these two currents must, it would seem, in the course of ages render it impossible for the lower strata to continue to supply the superior elements required to maintain the upper. If and when that stage is reached, national decay must set in.

In England, where the operation of the social ladder has been more effective and of longer duration than in any other country, there are indications that this stage is at hand. Our social ladder has provided and still provides a splendid array of talent, but already it has produced, as its complement, a large mass of very inefficient population. Foreign observers are constantly impressed with this; Mr. Price Collier,[1] for example, tells us that the million best of our population is the finest in the world; but that our lowest stratum is the most degraded and hopelessly inefficient.

Looking at the course of history widely, we may see, then, in the differentiation of social classes by the social ladder, and in the tendency of the upper strata to fail to reproduce themselves, an explanation of the cyclic course of civilisation. This has been ascribed by some authors[2] to race-crossing, followed by blending, and ultimately by stagnation consequent upon complete blending and the flowering period which coincides with it. But we now have a more adequate explanation of the decay which follows upon the blooming period. It is not mere stagnation, resulting from the achievement of social harmony and the relaxation of efforts at social adaptation and achieve-

[1] *England and the English.*

[2] Notably by Prof. Flinders Petrie in his *Revolutions of Civilisation.*

ment of all kinds. The decline is probably due as much, and perhaps in a much higher degree, to the exhaustion of the mass of the population, the completion of the draining process by which, throughout the whole period of the development of the cycle of civilisation, the best elements and strains have been drained off from the lower strata, brought to the top, and strained off.

It is interesting to speculate on the possible effect on this process of the fact that we are becoming more clearly conscious of these tendencies and subjecting them to scientific inquiry. Already the legislature has taken one small step of a eugenic nature and is soon to take another. The important thing is that we should recognise that men are not the helpless sport of blind forces, that mankind can control its own destiny in ever increasing degree as knowledge grows.

A word may be said in regard to sexual selection, which probably played a part in the evolution of the mental capacities of men. It would seem that, in the peoples among whom monogamy is the rule, it no longer operates to any appreciable degree. With the general excess of females, we could suppose that it still tended to race improvement only if the unmarried women were on the whole distinctly inferior to the married. But, if there is any difference, it is probably the other way; because the most able women are more and more attracted into independent careers. The further the so-called emancipation of women goes, the more will this be the case.

Civilisation, then, tends from the first to put an end to that elimination of the less fit individuals by the severities of Nature which we call natural selection; and, as soon as it has passed beyond its earliest stages, it brings to an end also the mortal conflicts of social groups and the consequent group selection, which was in all probability a main factor of racial progress in the prehistoric period. It abolishes also at an early stage the improving influence of sexual selection, which was probably the third principal condition of the development of the higher powers of mankind.

Civilisation replaces these modes of selection, which make for improvement of the racial qualities of peoples, by a number of modes of social selection, nearly all of which must have been, so far as we can see, negative or reversed selections—that is, selections making for deterioration of the mental qualities of the civilised peoples. In place of natural selection, group selection, and sexual selection, we have had at work, within each people in increasing degrees, various forms of social selection—military selection, selection by the towns, selection by the church, political selection with its exiles and its colonial system, and lastly economic selection, which has become exceedingly influential in recent years among ourselves. And all these, so far as can be seen, have operated mainly, among some peoples and in some ages very powerfully, to diminish the fertility of the best elements of the population and so to produce actual retrogression of the average intellectual capacity of peoples, and especially to deprive them of eugenic stocks, the stocks which were most fertile in individuals of exceptional capacity on whom the progress of civilisation and the relative power of nations chiefly depend.

M. de Lapouge's investigations of the matter have led him to a very melancholy conclusion. He attaches special importance to urban selection, as he calls it, in weeding out the best stocks. He writes—"There is no more agonising question than that of the exhaustion of our intellectual reserves by the influence of city-life. The public and our statesmen do not suspect it. But nevertheless it is the great danger of modern societies and especially of France. Of all the devastating influences which we have called social selections, selection by the town makes most powerfully for deterioration of peoples. Our towns are destroying all of the intelligent and energetic that have been spared us by the long centuries of disastrous selections. France has lost in the past almost all her dolicho-blond elements, and now are disappearing those of mixed stock and the best of the short-headed type. In all the continent of Europe the hour is at hand when there will remain only the inert and used up débris of our dead nations, pitiable remnants who will be the prey of unknown conquerors. Thus perished the Hellenic world, thus will perish the whole of our civilisation, if man does not make application of his knowledge of the principles of heredity, that tremendous power which to-day is bringing death and stagnation, but by the control of which science will enable us to secure safety and national vigour."[1]

It is possible that this conclusion gives too dark a picture of the tendencies of social selection in the civilised nations; but it does seem probable that with the advance of civilisation the tendency to reversed selection becomes strong.[2] We are at any rate compelled to conclude that it is impossible to discover evidence of any influences that can have made at all strongly for progressive evolution of intellectual capacity during the historic period; whereas a number of forms of selection seem to have worked against it and must at least have counterbalanced any factors making for improvement, and that therefore no advance has taken place in intellectual capacity, but more probably some deterioration has already occurred.

[1] *Op. cit.*, p. 407.

[2] De Lapouge does not stand alone in this opinion. Many biologists and leaders of thought have expressed it hardly less strongly, though not all of them have attached so much importance to the influence of the towns. It has been expressed in general terms by Dr. and Mrs. Whetham (in the *Hibbert Journal* for Oct., 1911),by Dean Inge in a number of forcible articles, by Mr. W. Bateson in his "Herbert Spencer Lecture" for 1912, and by other writers in a number of articles in the *Eugenics Review* and other journals.

The conclusion thus reached deductively is well borne out by the small amount of inductive evidence that is available. Such comparison as we can make between the leading modern nations and the civilised nations of antiquity tends rather to show that both as regards the average man, and as regards the intellectual endowment of exceptional men and the proportion of such men produced, the advantage lies with the ancient peoples. And the comparison of skull capacity or size of brain decidedly supports this conclusion. It has been found by a number of anthropologists that the average skull capacity of men of the late Stone Age in Europe was equal to, or greater than, that of modern Europeans. And in the main, on the large average, intellectual capacity varies with the size of the brain.

Our seeming intellectual superiority is a superiority of the traditional store of intellectual gains, a superiority of knowledge and of the instruments of the intellect, of language, and of the methods of mental operation by which knowledge is obtained, especially the mathematical and scientific methods in general.[1] Consider a single example frequently quoted to show the intellectual inferiority of the modern savage. It is said—Here is a poor savage who cannot count above ten without the help of his fingers and toes or other tallies; and we generally forget that we also should be incapable of counting above ten, had not our ancestors slowly devised the system of enumeration or verbal counting, and that, given such a system, the poor savage would be able to count as well as any of us.

[1] This conclusion may perhaps be said to be now generally accepted by those who have given any thought to the matter. A. R. Wallace argued strongly in this sense; the late Benjamin Kidd set out the evidence impressively in his *Social Evolution*, Chapter IX; and it is implied by all the many writers who, as we have noted, agree in regarding the processes of selection in the civilised nations as in the main reversed or detrimental.

The reader may be prepared to accept this conclusion as regards the intellectual capacities of mankind, and yet may be inclined to say—Surely the civilised peoples have progressed as regards their moral qualities throughout the historic period! Let us, therefore, consider this point separately for a moment.

Is there reason to believe that there has been progress of the innate moral disposition during the historic period? Here we are on still more difficult ground than when we considered the question of the progress of innate intellectual capacity.

The essence of the higher morality is the predominance of the altruistic motives over the egoistic, in the deliberately reasoned control of conduct. But morality in this sense is relatively rare in every age, and the great mass of moral conduct of men in general is the issue of mental processes of a simpler kind; it consists in doing what one believes to be right, in acting according to what one believes to be one's duty; no matter how that belief may have been arrived at. The tendency to do what one believes to be right, which for the vast majority of men has always been simply the tendency to conform to the code of morals accepted by his society, has an innate basis which may properly be called the social or moral disposition. At present I am not concerned to define the elements of our nature which make up the moral disposition.[1] The morality of a people, objectively considered, is the outcome of the interaction between their moral disposition, on the one hand, and the moral environment of the individuals, on the other; and the latter consists of two parts: (1) the traditional system of precepts, customs, laws, in short the code; (2) the traditional system of sanctions by which the code is upheld and enforced.

If we compare, in respect to this moral nature, the

[1] I refer the reader to my *Social Psychology*.

members of primitive societies with those of highly civilised societies, applying simply the criterion of conformity of conduct to the accepted code, we shall be impelled to the conclusion that the former, the savages and barbarians, have in general the moral nature much more highly developed than the members of civilised societies; for they conform on the whole very much more strictly to their moral codes. But such a conclusion would be hardly fair to the civilised peoples; first, because their social environment is more complex, so that the bearing of their moral code is less simple and direct; it is less easily obeyed, because its teachings are more generalised in form and do not provide clear irresistible rulings for all or any large proportion of the much greater variety of situations with which individuals find themselves confronted. Secondly, because the code is a higher one and makes greater demands upon the self-control of individuals. Thirdly, because not only is the code less clear and direct, but also the sanctions of conduct, civil and religious, are generally less obvious and immediate; and the effectiveness of both code and sanctions is weakened by the co-existence, within complex civilised societies, of more or less rival codes and systems of sanctions, which inevitably weaken the authority of one another; whereas the code and sanctions of the savage or barbarous society reign absolutely and without rivalry, so that men are not led to question their authority.

The conditions of moral conduct are, then, so different as to forbid any attempt to compare the innate moral dispositions of primitive and civilised peoples; and all we can do, in order to arrive at an opinion, is to consider whether the conditions have been such as to favour the evolution of the moral disposition, the innate basis of the social tendencies, during the nation-making period.

There can, I think, be no doubt that the principal condition of the evolution of the moral nature was group selection among primitive societies constantly at war with one another. In conflicts of that kind it must have been the solidarity of each group, resting upon the moral dispositions of individuals, the tendency of each individual to conform to the law and moral code of the society and to stand loyally by his leaders and comrades, which, more than anything else, determined success and survival in the struggle of the group for existence. At first, the nature of the code must have been of relatively small importance; the all-important condition of survival of the group must have been the strict obedience to it on the part of the members of the group.

This is not a deduction only from general principles. One may observe the effect of tribal conflict, on comparing, in various parts of the world, tribes that have long been subjected to its influence with closely allied tribes that have long led a peaceful existence.[1]

At a later stage, as the traditional codes of morality became differentiated and more complex with the increasing complexity of societies, the nature of these codes must have acquired an increasing influence in determining group survival; but it must still have been subordinate in importance to the degree of development of the moral disposition; for a society with an inferior moral code, strictly conformed to by its members, would in the long run have better chances of survival than one with a higher code less strictly observed. Hence, the higher more difficult codes could only be attained by those peoples among whom the instinctive basis of social conduct had become highly evolved by a long process of group selection.

But, on passing into the stage of settled societies of large extent, that is to say, as peoples passed from the stage of tribal organisation to that of national organisation, the

[1] In this connection I may again refer to *The Pagan Tribes of Borneo*, by C. Hose and W. McDougall.

evolution of the social disposition through the mortal conflict of groups must have tended to come to an end; because group selection became less active, the conflicts between the larger and less numerous societies or groups became rarer and also less fatal to the vanquished societies. In other words, during the historic period failure in conflict has not usually meant extermination; national cultures and the power and glory of nations have come and gone, but the various peoples, the units of conflict, have in the main survived their failures and persisted in living. Group selection, the main condition of evolution of the social disposition, has, therefore, been abolished; and of the various forms of social selection operating within societies, the chief of which we have briefly noticed, no one seems to have been of a nature to produce further evolution of the social disposition; all of them must rather have operated adversely to it. Military selection, selection by the Church's rule of celibacy, political selection—all these must have fallen most heavily on the individuals in whom the social disposition was strong, whose conduct was influenced largely by the sense of duty, and less by the individual impulses and desires.

We may conclude, then, with some confidence that there has not been further evolution of the innate moral disposition in the historic period. This conclusion is greatly at variance with popular conceptions; we are apt to pride ourselves upon our superior morality; to point to our humanitarian laws and institutions, to our tenderness for the weak, the poor, and the suffering; to our regard even for the welfare of savage peoples, whom we no longer deliberately exterminate, and for domestic animals; and to suppose that all this shows modern civilised men to be innately superior in morality to their ancestors and to the barbarous peoples. But our conclusion that the difference implies merely an evolution of moral tradition, not of moral nature, will appear probable if we reflect upon the fact that a widespread change of this kind in respect to some department of conduct has sometimes been produced within a very short space of time, even within the lifetime of one generation. Take the attitude of Englishmen towards slavery and the African slave trade. It is hardly more than half a century since large numbers of Englishmen, or men of English origin, owned great gangs of slaves or drew their wealth from slave labour; yet now most of us look with horror upon slavery of every kind. Take the case of kindness to domestic animals. It is a comparatively recent tradition; and, within the memories of those who are not yet middle-aged, a great improvement has taken place. Again, there are many persons who, while tender to their domestic animals, are entirely brutal where wild animals are concerned, since public opinion or traditional morality does not yet bear so strongly upon our relations to them. Again, it is not long since in our factories, our prisons, our schools, the most horrible tortures were applied to our fellow citizens without provoking any protest; while now we display perhaps an excessive tenderness and have passed law after law to protect the feeble against the strong.

The mental development of peoples in the historic period has, therefore, not consisted in, nor been caused by, nor in all probability has it been accompanied by, any appreciable evolution of innate intellectual or moral capacities beyond the degrees achieved in the race-making period, before the modern nations began to take shape. There is no reason to think that we are intellectually or morally superior by nature to our savage ancestors. Such superiority of morals and intellectual power as we enjoy has resulted from the improvement and extension of the intellectual and moral traditions and the accompanying evolution of social organisation.

A different conclusion was reached by the late Benjamin Kidd in his *Social Evolution*, which has enjoyed a very wide circulation,[1] and it seems worth while therefore to examine very briefly the author's position. Mr. Kidd saw clearly and argued convincingly that the innate intellectual capacities have not improved during the historic period; but he held that the innate moral tendencies have been greatly improved during this period; or rather he distinguished between the innate moral tendencies and the innate religious tendencies; and, while rejecting Herbert Spencer's view that the moral tendencies (as thus arbitrarily distinguished from the religious tendencies) are slowly becoming improved and strengthened in the civilised peoples, he held that the innate religious tendencies are being greatly improved and strengthened; and he regarded this as the underlying condition of all "social evolution." In support of his view he cited an impressive array of facts illustrating the general softening of manners and morals among the civilised peoples, especially the legislative changes which have given political power to the masses of the people. That these evidences of a general softening of manners and a great extension of social sympathy are very striking we must all agree; but Kidd advanced no serious argument in favour of his contention that these changes have been due to some change or improvement of the innate qualities of the peoples among whom they have appeared. And he did not suggest any way in which this alleged improvement or accentuation of the innate religious tendencies may have been brought about. He attributed it wholly to the influence of the Christian religion. Now, if Kidd had accepted the Lamarckian principle of the transmission of acquired tendencies or effects of use and habit, he might reasonably

[1] It has been translated into nine languages and was reprinted ten times in the first year after its publication.

have attributed the alleged improvement to such influence. But he sternly rejected that principle and proclaimed himself a rigid exponent of the Neo-Darwinian school, which attributes all racial changes to selection. He even assumed the truth of the doctrine that, in the absence of selective processes making for its improvement, every race must inevitably degenerate. It might, then, have been expected that he would have attempted to show how Christianity can be supposed to have favoured the improvement by selection of the innate religious tendencies. Yet he made no attempt in this direction. He seems to have been aware that his view encounters a great difficulty in the fact that Christianity powerfully swayed the peoples of Europe for many centuries during which little or no progress in civilisation was effected, whereas rapid and accelerating progress of many kinds has marked the last three centuries. He sought to meet this difficulty by attributing the rapid progress of recent centuries to the influence of the Protestant form of Christianity, alleging that it promotes the evolution of the religious tendencies more powerfully than other forms. Yet this view of the matter, even if it were acceptable, would leave the Reformation itself quite unexplained. Kidd seems to hint that, throughout the earlier centuries of the dominance of the Christian religion in Europe, it was slowly effecting the alleged improvement of the religious tendencies in the mass of the people, without these being able to manifest themselves in social life, until they somehow broke loose at the time of the Reformation and began for the first time to operate on a great scale and with tremendous force. The view might have some plausibility coming from the mouth of a disciple of Lamarck, but it cannot be reconciled with Kidd's strictly Neo-Darwinian principles. There is, then, nothing in Kidd's grandiloquent and loosely reasoned but always interesting pages, to justify any belief in the improvement of the innate moral disposition during the historic period.[1]

Before leaving this difficult question of the extent and nature of changes in the innate qualities of peoples during the historic period, I would define in the following way the position that seems to me to be well founded. There have been no considerable changes of innate qualities; and what changes have occurred have probably been of the nature of retrogression, rather than of advance or improvement; and this is true of both intellectual and moral qualities. The improvements of civilised peoples are wholly improvements of the intellectual and moral traditions. All the great and obvious changes of social life are in the main changes of these traditions. Nevertheless, such differences of innate qualities as exist between the different peoples are very important, because of their cumulative influence upon their traditions. And, especially, the

[1] Shortly before his death Mr. Kidd published (in the year 1918) his *Science of Power*. In this book he showed a complete change of face on the question of the importance of innate qualities. He denied all importance to changes of innate qualities, whether for better or worse, because, as he maintained, "the social heredity transmitted through social culture is infinitely more important to a people than any heredity inborn in the individual thereof" (p. 273); and he made in this book a violent and scornful attack upon the late Francis Galton and upon all who follow him in believing that the decay or improvement of the racial qualities of a people are of importance for its prosperity and development, and who, therefore, approve of Galton's effort to found a science of Eugenics. Kidd did not anywhere in his last book acknowledge that he had made this very great change of principle, which completely undermines the whole argument of his *Social Evolution*, but complacently suggested that, as Newton and Darwin are regarded as the fathers of modern physical and biological science respectively, so in the future Kidd will be regarded as having founded anew in his *Social Evolution* the science of society. On reading the *Science of Power* after having written this chapter, I was amazed at this assumption on behalf of a book whose most fundamental doctrine the author had himself renounced, and I turned again to the earlier work to verify my brief summary of its argument. I confess that it is not easy to make sure of what the author was driving at. But I find that Kidd, in discussing the influence

innate superiorities of the leading peoples, though relatively small, are of essential significance; and it is of the first importance for the future prosperity of the great nations of the present time that they should not suffer any deterioration of their innate qualities; for they alone have attained just such a level of innate excellence as renders possible the existence of civilisation and the growth and continued progress of great nations. Especially is it essential that they should continue to produce in large numbers those persons of exceptional moral and intellectual endowments, whose influence alone can maintain the vitality of the national traditions and who alone can add anything of value to them.

---

of religious systems, wrote (on p. 307): "Natural selection seems, in short, to be steadily evolving in the race that type of character upon which these forces act most readily and efficiently; that is to say, it is evolving religious character in the first instance, and intellectual character only as a secondary product in association with it." On the following page I find—"The race would, in fact, appear to be growing more and more religious," and "a preponderating element in the type of character which the evolutionary forces at work in human society are slowly developing, would appear to be the sense of reverence." And there are many other passages which, in spite of the habitual lack of precision of Kidd's language, can only be interpreted to mean that the improvement of moral or religious character, on which he so strongly insists as a feature of recent centuries, involves and depends upon improvement of innate qualities in the mass of the people.

CHAPTER XIX

The Progress of Nations in their Youth

WE have found reason to believe that during the historic period the peoples of Europe have made no progress in innate qualities, moral or intellectual; yet that period has been characterised by immense mental development, a development essentially of the collective mind. The most striking result of the formation of nations and the development of civilisation has been this replacement of the progress of the individual mind by the progress of the collective national mind. And the most interesting and important problem of group psychology is—What are the conditions of the progressive development of the collective mind?

I insist that this is distinctly and primarily a psychological problem. The conclusion we have just reached, to the effect that it is not produced by and does not imply a racial evolution, shows that it is not to be regarded as a biological problem. It cannot be treated as a problem of economics or of politics; these sciences only touch its fringe at special points.

We have before us the significant fact that in some cases the collective mind of a nation has remained stationary at a rudimentary stage of development for long ages; while in other nations the collective mind has developed at a constantly accelerating rate, becoming more highly differentiated and specialised and at the same time more highly integrated, has in fact developed in a way closely analogous to the evolution of the individual mind. The collective mind, in thus developing, reacts upon the development of individual minds, raising all far above the level they could independently attain, and some in each generation to a very high level both intellectually and morally.

The merest outline of a discussion of this great problem is alone possible. I can do no more than offer some suggestions toward the full solution of it. Let us note, first, that continued progress, far from being the rule, as is commonly assumed by popular writers, has been a rare exception, as Sir H. Maine pointed out in *Ancient Law*. He wrote—"In spite of overwhelming evidence, it is most difficult for a citizen of Western Europe to bring thoroughly home to himself the truth that the civilisation which surrounds him is a rare exception in the history of the world." "It is indisputable that much the greatest part of mankind has never shown a particle of desire that its civil institutions should be improved, since the moment when external completeness was first given to their embodiment in some permanent record. Except in a small section of the world, there has been nothing like the gradual amelioration of a legal system." And what is true of systems of law is true of all the other elements of the intellectual and moral tradition which constitute a civilisation or national culture.

Sir H. Maine added—"The difference between the stationary and progressive societies is, however, one of the great secrets which inquiry has yet to penetrate." His own contribution, which he regarded as a partial solution only, was that the difference depends in part upon the period at which the customs of a people become codified in written law. If, as the tribes of a people become settled and enter upon a national existence, there is no written code of law and custom, customs, he urged, which at their origin were socially advantageous tend to become extended by analogy to other fields of practice and to assume an excessive and senseless rigour; for example, the custom of cleanliness becomes the exceedingly elaborate ritual of purification, which among the Hindus limits and restrains social life at every point. Or a useful distinction of classes becomes a rigid caste system, than which nothing is more prejudicial to progress, intellectual or moral. The continuation of the process of extension by analogy through long ages has resulted in nearly all the uncivilised and less civilised peoples of the modern world being bound down on every hand by a system of rigid and worse than useless customs, which, restricting both thought and action, render progress impossible. On the other hand, early codification of custom in a system of written laws secures that thereafter custom shall not develop in this blind unintelligent and socially prejudicial manner, but shall be developed only by deliberate intention and the reasoned forethought of the ruling powers of society; it will then develop in the main, in spite of many mistakes, in a way which promotes the efficiency of social life and the welfare of society.

Maine's suggestion is in harmony with the fact that the progressive peoples have not been those who invented or learnt the art of writing at an early period. Writing and the written codification of customary law could not be invented by any people until they had attained to a settled life and a considerable degree of social organisation; and then, when the invention was worked out sufficiently, the damage had been done, socially advantageous customs had already degenerated into useless rites and ceremonial observances; and writing served only to establish these more firmly, to fix their yoke upon the necks of the people, as in the case of the Hindus.

On the other hand, the progressive peoples have been

those who remained in a savage or barbarous condition until a relatively late period, and who then acquired by imitation the arts of writing and of reducing custom to written law, acquired them in a fully developed condition from the peoples who had invented and developed these arts. They have, therefore, enjoyed the advantages of written laws from the beginning of their civilisation.

But, as Maine recognised, the acquisition of writing at the outset of national life is by no means sufficient to account for the progressiveness of the nations of South and Western Europe; we must seek other causes and conditions of their mental progress.

We have already noted certain features of the racial constitution which were probably essential to the continued progress of the European peoples—namely, the high degree of evolution of the social disposition through group selection in the long prehistoric or race-making period; a group selection which probably was far more severe and prolonged than the peoples of any other part of the world were subjected to; and which in turn was due probably, as we have seen, to the great diversity of physical surroundings and to the comparative severity of the climate of Europe, especially of the northerly parts in which the most progressive European race was formed; for these physical conditions generated in the race an innate energy, a capacity for sustained effort.

Without the highly developed social disposition in the mass of their members, primitive societies could not have survived those changes of custom and institution which were essential features of their progress. Without their innate energy, active rivalry and competition, which have been chief factors in social progress, would not have been constant features of the relations of these societies. Still the possession of a highly evolved social disposition by the European peoples does not in itself suffice to account for the continued mental evolution of the leading nations. For not all the European peoples have progressed; and, of those that have progressed, some have done so much more effectively than others.

Let us first examine the question—In what has progress primarily consisted? Has it been primarily a progress of the moral or of the intellectual traditions? As I mentioned in an earlier chapter, we have here one of the main points of dispute.

Buckle was the great advocate of the primacy of intellectual development in the sense of increase of natural knowledge. The argument by which he sought to establish the position runs as follows: Progress must have been due to improvement either of moral or of intellectual principles. But moral principles have been almost the same in all ages. "To do good to others; to love your neighbour as yourself; to honour your parents—these and a few others, are the sole essentials of morals; but they have been known for thousands of years, and not one jot or tittle has been added to them by all the sermons, homilies, and text-books which moralists and theologians have been able to produce." On the other hand, knowledge and intellectual principles have made immense strides; hence all progress must have been primarily intellectual rather than moral. Buckle did not deny that there has been some moral progress; rather he insisted upon it as an essential feature of the progress of civilisation; but it has, according to him, consisted only in the more effective operation of unchanging principles, and this more effective operation is secondary to, and due to, intellectual progress.

I think we must agree with Buckle that the increasing store of knowledge and the increased command over nature that comes with it has been the primary condition of the progress of nations. For, since the early middle ages, the moral natures of men and the teaching of Christianity have been the same in all essentials; yet for many centuries there was practically no progress. Kidd himself admitted that progress only set in rapidly about the time of the Reformation. And it is notorious that this progress, including the Reformation itself, was due to the stimulation of the intellect by a number of influences—by the renewed study of classical art and literature, by the discovery of the New World, by the increased intercourse of nations resulting from the improvement of the art of navigation, by the accumulation of wealth and the formation of a powerful middle class. It is clear also that religion, far from having been the sufficient cause or instrument of progress, was largely responsible for the stagnation of the middle ages, through sternly repressing the sceptical spirit and leading off men's minds from inquiry into natural laws, to the discussion of many topics on which it was impossible to achieve knowledge and which were necessarily barren of results making for human progress. Nevertheless the Christian religion has in the long run co-operated in forwarding the mental evolution of the European peoples in an important manner which we must briefly consider later—namely, through its effects upon social organisation.

Without raising the question of the natural or supernatural origin of religion or of any particular religion, we may say that from the point of view of national life, a religion is essentially a system of supernatural sanctions for social conduct, for conduct conforming to the moral code of the society, and especially for customs regulating the family and the relations of the sexes, on which, more than on anything else, social stability depends. It is, thus, the great conservative agency; for it enforces the observance of custom by a system of rewards and punishments; in the earlier stages of society, especially by punishments. It is essentially intolerant of change of custom or belief; and even the Christian religion has exemplified this principle in the terrible persecutions and innumerable wars for which it has been responsible.

The great function and tendency of any religion, once established among a people, is to preserve intact the current moral code and to secure conformity to it. Nevertheless, some religions are less prohibitive of progress than others; and, when such a religion replaces a more restrictive one, an important condition of progress is realised. But, in so far as progress is then favoured, this is not due to the changed operations of the religious emotions and sentiments; it is due to the great religious teachers who have succeeded in breaking down the bonds imposed by the more primitive religion, and so have given freer play to the intellectual faculties; the improvements of religious systems have been negative or permissive conditions of progress, rather than its effective cause.

Progress has, then, always resulted primarily from the gains made by the intellect and added to the intellectual tradition, that is to say, from the progress of knowledge. Nevertheless, the free play of the intellect is always a danger to society, for the reason that the customs and moral code of a society, however imperfect and sanctioned by a religion however narrow, are yet the bonds by which alone it can be held together; to their influence has been largely due in every age the subordination by the members of a society of their individual egoistic ends to the welfare of the society as a whole.

The spirit of inquiry, which always leads men to question the authority of these customs and moral codes and of their religious sanctions and thus tends to weaken them, is, then, a socially disruptive force, at the same time that it is the source of all progress. Hence, though the free play of the intellect and of the spirit of inquiry may secure for a time the rapid progress of civilisation, it cannot alone secure continued progress. Continued progress has only resulted where there has been maintained a happy balance between the conservative and the progressive forces; between the authority of custom and the moral code on the one hand and the free activity of the intellect on the other. Wherever the progressive force has outrun the conservative, progress has been first rapid and then has come abruptly to an end. Greece exemplified this process in the clearest manner. It was the excessive seeking of individuals for their own power and glory, unrestrained by the customs and religious systems which their intellect had outgrown, that ruptured the bonds of society, plunged the State into war and civil strife, and eventually destroyed it by the extermination of the Greek aristocracy. The same is true of the brilliant but brief periods of rapid progress exhibited by the mediæval Italian States. Intellect outran and undermined morals, and progress was brought to an end.

Some observers have maintained that history will pass the same verdict upon modern France, and that most of our leading nations of the present day are seriously threatened by the same danger.

Any long continued progressive evolution of the mind of a people has been, then, a rare exception in the history of the world; partly because the free play of the spirit of inquiry and of the intellectual faculties, which is the source of all progress, exerts a socially disruptive tendency, so that progress is by its very nature dangerous to the stability of any nation; but partly also because the free play of the spirit of inquiry has been so rarely achieved or permitted, so that even such progress as has led on to social disruption has been exceptional.

A long period of intellectual and moral stagnation in the rigid bonds of custom and religion has been the rule for nearly all the peoples of the earth, so soon as they had attained to a settled mode of existence. The primary question, then, to be answered in seeking to account for the progress of nations, is—What conditions enabled the spirit of inquiry to break the bonds of custom and religion and to extend man's knowledge of man and of the world in which he lives?

Bagehot, in considering more particularly the progress of political institutions, put the problem in much the same way. He pointed out that the first age of the life of nations is always an age dominated by custom resting on unquestioned religious sanctions; an age in which there is often a vast amount of discussion of detail, as, for example, discussion of the details of military expeditions, but never discussion of principles; and he maintained that an age marked by the discussion of principles, involving the questioning of traditions, moral and intellectual, initiates and characterises every period of progress.

There is much to be said for the view that the most important condition of progress in its earlier stages was in most cases, perhaps in all, the conquest of a more primitive people by one more advanced in culture or of superior racial type, who remained to settle in the conquered territory, and, not driving out or exterminating the conquered inhabitants, established themselves as a governing class. History and archæology show that this occurred at least once in most of the areas where nations have developed spontaneously to any considerable degree; the earliest known instances being those of Egypt and Chaldea as long as ten thousand years ago. The same thing occurred again in India, and later still in Greece; and throughout early European history the process was frequently repeated in various areas. Every one of the modern peoples of Europe has been formed through such fusion by conquest of two peoples, in some instances several times repeated; and, though none of these modern European peoples originated their own civilisations, but largely took over by imitation the civilisation ready made for them by the more precocious peoples of Asia and by Greece and Rome, these fusions and the resultant composite character of the European peoples no doubt have tended greatly to promote progress. And it is easy to see how in several ways such a fusion by conquest of two peoples must have tended to set free the spirit of inquiry, that prime condition of progress. Three of these seem to have been of chief importance.

The most obvious way in which progress has been promoted was that the conquering invaders became a leisured aristocracy, having their material needs supplied by the labour of the indigenous population, which became a more or less servile class. All the ancient civilisations were thus founded upon servile labour. We may be sure that, until such a social system resulted from conquest, no people made much progress; because all individuals were fully occupied in securing their means of subsistence, either by warfare, by the tending of herds, or by agriculture. Each people was self-supporting, and knew no or few needs beyond those which their own labour was able to supply; and labour was individual, or was co-operative only among small groups, such as the communal family groups. It could, therefore, undertake no great works, whether of building or engineering, such as large public buildings, irrigation, or road making. Each family consumed what it produced, and consequently there was no large accumulation of capital; for there were no motives for storing up their primitive wealth, and generally no wealth of durable and storable form.

But, as soon as a ruling class could dispose of the labour of a large part of the population, making them work for a mere subsistence wage, there was initiated that regime of capital and labour on which, up to the present time, all civilisation has been founded. Wealth was accumulated; great works, such as the pyramids, demanding enormous expenditure of human life and work, could be undertaken; and a leisured class was created, which, being freed from the necessity of bodily toil, was able to turn its energy to speculative inquiry, to the enjoyment of art and luxury, to directing and organising the labour of the multitude, to inventing the tools that render labour more effective, to studying natural phenomena such as the cycle of the seasons, a more accurate knowledge of which added to the productivity of labour; for it was in the service of wealth production, that in the main science arose, especially mathematics and mechanical and astronomical science; arithmetic and geometry through the need of a practical art of measurement, astronomical science through the need of foreseeing the seasons.

The desire to enjoy art and luxury is one which feeds itself and grows, when once aroused; and it was these growing desires of the leisured and wealthy classes which created trade, or at any rate first developed it beyond the merest rudiments; and in doing so led to regular and friendly intercourse between nations.

A second very important result of such fusion by conquest must have been the breaking up to some extent of custom and the weakening of the religious sanctions. Under the new regime, both the conquering and the conquered peoples would find their old customs unsuited to their novel social relations, and inadequate to regulate their changed occupations. The old customs of both would inevitably be thrown into the melting pot; at the same time, the religious sanctions of both would be weakened by the intimate contact of two systems, neither of which, in the presence of a rival system, would henceforth be able to claim unquestioned authority, until one had suppressed the other or a stable synthesis of the two had been effected. So long as each individual never had intercourse with any but those who accepted the national or tribal religion, it was well-nigh impossible for anyone to question its authority; but as soon as the devotees of two religions lived intermixed, the question—Which religion was true? must inevitably have arisen in some minds. The weight of custom and of religious sanctions, which lies so heavily on a primitive society, restricting all enterprise, forbidding inquiry and repressing the use of the intellectual powers, would thus be lightened and scope be given for experiment in thought and action. And either people, coming into more or less intimate contact for the first time with a system of beliefs and customs and institutions other than their own, must have been led to compare, discuss, and reflect upon these things; the sceptical spirit and the intellect must have been greatly stimulated. There must have been a conflict of ideas and the initiation of an age of discussion. In short such a fusion by conquest must have broken up what Bagehot calls the "cake of custom" as nothing else could, and so have rendered the intellectual and moral traditions once more plastic and capable of progress.

No doubt in many cases such disintegration of the old systems went too far, and the society, before it could evolve anew a sufficiently strong and adequate system of customs and sanctions, went to pieces. In modern times many primitive societies have been broken up and destroyed in just this way—namely, their customs and the religious sanctions of their morality have been undermined and weakened by the contact of the more complex systems of civilised men, and they have not been able to assimilate the new system rapidly enough to enable it effectively to replace their own shaken and decaying code.

A third way in which the fusion by conquest of two peoples must have made for progress was by biological blending, the crossing by intermarriage of the two stocks. We have seen that there is a considerable amount of evidence to show that, when two stocks are very widely different in mental and physical characters, the result of crossing is likely to be bad; the crossed race is likely to be inferior to, and less fit for the battle of life than, both parental stocks; the characters of individuals will be apt to be made up of a number of elements more or less inconsistent with one another; such a composite character made up of inharmonious elements will be apt to be unstable and constantly at war with itself. Character of this kind and the tragic struggles to which it is liable to find itself committed has been well described in fiction by a number of authors, especially in stories of the Mulattoes of America. On the physical side it has been shown that such cross-bred races tend to die out owing to lack of balance of the physical constitution.

On the other hand, we saw that the crossing of two closely allied racial stocks seems to have a tendency to produce a cross-bred race superior to both parent stocks, and especially to produce a variable stock. It is, I think, probable that the frequently repeated blending of allied stocks in Europe has been the fundamental biological condition of the capacity of the European peoples for progressive national life.

In the case of the conquest of one people by another differing very markedly in racial qualities, there seem to be two alternatives equally prejudicial to the continued progress of the nation so formed. On the one hand, free intermarriage may take place, resulting in an inferior cross-bred race incapable of high civilisation, as seems to have occurred in most of the countries of South America, where it is with the greatest difficulty that the outward forms of the high civilisation which they have imitated from Europe are maintained. On the other hand, where especially the outward physical characters are very different, the conquering people may hold itself apart from the conquered, and maintain itself as a ruling class, which prides itself on the purity of its blood and which tends to harden into a caste. Such conquest without subsequent blending gives rise to a civilisation which, being founded upon a rigid caste system, is incapable of continued progress. This is what has happened in India. The fair-skinned Aryan invaders despised the dark-skinned indigenous peoples, whom they spoke of as being scarcely human; and, in spite of a good deal of crossing, they have in theory and, in the case of the Brahmans, at least, to a considerable extent, in practice, maintained the purity of their blood, by means of the development of the caste system.

Europe on the other hand was fortunate in that all the different peoples, or most of the peoples, from which its nations have been formed were of allied race; they were all, with few exceptions, of the white race, sufficiently nearly allied not to produce inferior cross-races but rather to produce some superior sub-races. The conquered peoples have been so similar to their conquerors in physical type that crossing could take place without the cross-bred offspring bearing the indelible marks of inferior or mixed parentage, such as a dark skin or a woolly head. Hence, although caste systems were formed, they did not prove rigid; free intermarriage took place, and it was not impossible for individuals of the conquered race or of the mixed stock to rise into the superior ruling class. The importance of this may be seen, on reflecting how the merest trace of Negro blood in individuals of mixed origin in North America is apt to show itself in the physical features and how, even in that enlightened and Christian country, a trace so revealed suffices to condemn a man, no matter how great his powers or refined his character, to remain a member of the inferior caste.

But, apart from the possible improvement of the racial qualities of the whole people, or of the average individuals in general, which may well have occurred in Europe, the biological blending of allied races may give important advantages to the resulting people in another way—namely, by increasing its variability, the variability of its mental qualities. If a people is extremely homogeneous in the racial sense, it may be expected to display little variability, its members will be of essentially similar mental qualities and of a uniform level of mental capacity; and this will tend to make them a very stable, but a very conservative unprogressive, nation. This seems to be true of China, and to be in large part the source of its extreme stability and extreme conservatism.

Where, on the other hand, a people is formed by the intimate blending by intermarriage of two or more racial stocks, it is likely to be a variable one; there will be large departures in many directions from the average type of mental ability; and there will be individuals varying by excess of development of various capacities, as well as others varying by defect of development.

And a people of variable and therefore widely diversified mental capacities will, even though its average capacity is no greater than that of a more homogeneous people, be more likely to make progress in civilisation; and this for three reasons.

First, variability is the essential condition of all race progress by biological adaptation; for it is by the selection of variations, the survival and multiplication of types varying in certain directions in larger proportions than the average type, that all race progress and adaptation seems to have been achieved. Hence, increased variability, resulting from the blending of races, will render a people so formed capable of race progress and of more rapid adaptation; for example, in the peoples of Northern Europe, it would have favoured the adaptation of the constitutions of the people to the severity of the climate, and to those peculiar social conditions which, as we have seen reason to believe, have been the source of their unique combination of qualities.

Secondly, variability of mental qualities would be favourable to the coming of the age of discussion; for in such a people custom would rule with less force, its sway would be more apt to be questioned and disputed, than among a highly homogeneous people.

Thirdly, and this is probably the most important manner in which race blending has favoured the progress of nations, among the variations from the average type produced by race crossing would be men of exceptional capacities in various directions.

We have already noted that all progress of the intellectual and moral traditions eventually depends upon the activities of men of exceptional powers of various kinds, upon the great religious or ethical teacher, the inventor, the artist, the discoverer. A people may, like the Chinese, have a high average capacity of intellectual ability; but, if it cannot from time to time produce men of far more than average capacity along various lines, it will not progress very far spontaneously. Exceptional intellectual capacity is, however, a variation from the type, as the biologists say, just such as may be expected to result from race blending; there will be, among the variations in all directions, variations in the direction of exceptional capacity of various kinds. Hence a nation of blended variable stocks will, other things being the same, be far more likely to be capable of continued evolution than a homogeneous people of equal average mental capacity, among whom few men are capable of rising to any distinguished height.

This view of the effects of race blending is borne out empirically by the comparison of the peoples of the world. The European peoples have been the most progressive, and they, more than all others, have been formed by repeated blendings of allied stocks. Within Europe it is the peoples among whom this blending has been carried furthest who have proved most progressive—the French, the English, and the Italian; and, conversely, the least blended peoples have been the most backward, and have contributed least to the general progress of civilisation in Europe; for example the large, almost purely Slav, population which forms the bulk of the Russian nation.

We pass on to consider other conditions which have contributed to setting free and stimulating the spirit of inquiry. We have seen that physical environment played a predominant part in moulding the mental qualities of races in the prehistoric period. And we must recognise that, although with the beginning of settled national life it probably ceased to modify race qualities to any considerable extent, it has yet been important in favouring the rapid evolution of the intellectual tradition of some peoples, and this in several ways. First, by its direct influence upon the minds of individuals. Buckle and others have pointed out that, while, in India and throughout a great part of Asia, the physical environment was unfavourable to intellectual progress, while its vast and terrible aspects fertilised the superstition of the people, and repressed the spirit of inquiry by rendering hopeless any attempt to cope with its terrific displays of force, in Europe, and especially in South and Western Europe, the comparatively small scale on which the physical features are planned and the relative feebleness of the forces of nature encouraged men to adopt a bolder attitude towards them.

Buckle, contrasting Greece with India in this respect, showed how the physical features of both countries were reflected in their national cultures; how, while the Hindus cringed in fear before monstrous and cruel gods, the Greeks fashioned their gods in their own image, simply personifying each leading human attribute, and made of them a genial family of beings, differing from men and women in little but their immortality and their superior facilities for the enjoyment of life. In general the buoyancy and serenity of the Greek attitude towards life and nature reflected the beautiful, secure and diversified aspects of their physical environment. In such an atmosphere the spirit of inquiry would naturally flourish more freely than where man's spirit was oppressed by the fear of terrible and uncontrollable forces, and where he was made to feel too keenly the limitations of his mental and physical powers. Buckle summed up his review of these effects as follows: "In the civilisations exterior to Europe, all nature conspired to increase the authority of the imaginative faculties and to weaken the authority of the reasoning ones. In Europe has operated a law the reverse of this, by virtue of which the tendency of natural phenomena is, on the whole, to limit the imagination, and embolden the understanding; thus inspiriting man with confidence in his own resources, and facilitating the increase of his knowledge, by encouraging that bold, inquisitive and scientific spirit, which is constantly advancing and on which all future progress must depend."

I think we must accept this view of the importance of the direct action of physical environment on the minds of individuals. To deny, as Hegel did, the important influence of physical environment upon the development of Greek culture, because the Turks have enjoyed a similar climate without producing a similar culture, is unreasonable. The progress of civilisation has always been the result of a multiplicity of causes and conditions; and we cannot deny all importance to any one, whether race or climate or social organisation or religion or any other, because in some particular instance it has failed to produce the progress of which in other instances it has been one of a number of co-operating causes.

The diversity and small scale of the physical features of South and Western Europe has favoured the progress of the intellectual tradition in another important way. The land is divided by natural barriers into a number of natural territories, the population of each of which has naturally tended to become one nation and to develop a national culture. In this way there arose a number of nations and States in close proximity with one another, yet each developing along its own lines. When the development of wealth and commerce brought these diversified cultures into friendly intercourse with one another, the exchange of ideas and the general imitation of the useful arts of one people by its neighbours must have made very strongly for progress; the culture of each of a group of neighbouring peoples no longer progressed only by the addition of the ideas and inventions of its own exceptional intellects; but each group had the opportunity of selecting and imitatively adopting whatever seemed to them best among the ideas, the arts, and inventions of the neighbouring peoples.

It is generally admitted that this was one of the main conditions of the rapid development of the culture of the ancient Greeks, situated as they were within easy reach of several of the oldest civilisations, those of Egypt and of South-Eastern Asia; they were also within reach of a number of less civilised peoples, and therefore enjoyed opportunities for trade of a kind which, being peculiarly lucrative, has in all ages hastened the acquisition of wealth and capital and stimulated the development of commerce.

All the most progressive European peoples have enjoyed similar advantages; and it has been maintained with some plausibility that the principal cause of the shifting of the centre of progressive civilisation from the Eastern Mediterranean to the west of Europe has been the improvement of the art of navigation and the discovery of the New World and of the sea route to Asia and the East Indies; for these gave the western countries the most advantageous positions for the conduct of a world wide commerce. No doubt the factor mentioned has been important in producing this change.

But, when we consider the ancient European civilisations and compare them with our own, we realise that, in spite of all the circumstances which we have enumerated and briefly considered as factors stimulating the spirit of inquiry and making for progress of their intellectual tradition, and in spite of their brilliant and in some respects unapproachable achievements, they were nevertheless radically incapable of continued progress. Greek civilisation certainly progressed at a marvellous rate for some centuries; yet there is every reason to believe that it bore within itself the inevitable causes of its ultimate decay or stagnation. And, when we consider Roman civilisation, we see that, through all the long centuries of the greatness of Rome, it was essentially unprogressive. There was no continued evolution of the national mind and character. Save in respect to the single province of law, Roman civilisation, when it entered upon the period of its decay, had not appreciably progressed in any essential respects beyond the stage reached more than a thousand years earlier. Rome was in fact less truly a nation in its later than in its earlier age. It had superficially imitated rather more of Greek culture, and it had incorporated a number of bizarre elements from the many peoples which had been brought under the sway of the Roman sceptre; but neither in religion, nor in philosophy, nor art, nor science, nor in any of the practical modes of controlling the forces of nature, had it made any substantial gains; and its social organisation tended more and more to the type of a centralised irresponsible bureaucracy.[1]

On the other hand, it cannot be denied that in the last thousand years the nations of Western Europe have made immense progress; nor that this progress has been accelerating from century to century in a way which seemed to reach a climax in the wonderful century just closed; though there appears good prospect of continued progress and perhaps of continued acceleration throughout the century to come and perhaps for many more.

What then is the cause of this great difference between the civilisation of Western Europe and all preceding civilisations? The difference is, I think, essentially due to difference of social organisation. As argued in a previous chapter, social organisation was of less influence in the earlier ages, but has assumed a constantly increasing

[1] Otto Seeck (*op. cit.*, vol. i., p. 270) writes—"The equipment of the legionaries remained unchanged from Augustus to Diocletian: no improvements of tactics, no new munitions of war were brought into use during more than three hundred years. The Roman saw his enemies becoming ever more terrible, his own army ever less efficient; for now this, now that, Province was laid waste and all were threatened. It was, therefore to the most urgent interest of every citizen that this state of affairs should be remedied; the most cultured circles were familiar with the needs of the army, for all the higher officers came from the class of Senators and nobles. Nevertheless, there appeared not a single invention, which might have assured to the Roman soldiers their erstwhile superiority! Books indeed were written upon tactics, strategy and fortification, but their authors almost without exception were content to expound in a formal manner what their more capable forefathers had taught; in this literature the expression of any new idea was carefully avoided. . . . As in the military sphere, so also was it in all others. Neither in agriculture, nor in handicrafts, nor in the practice of statecraft, did a new idea of any importance appear since the first century after Christ. Literature and art also moved only in sterile imitation, which became always more poverty-stricken and technically feebler."

importance throughout the evolution of civilisation; and it is now predominant over all other conditions. We must, then, first define this difference of social organisation; secondly, we must show how it makes for progress; and thirdly, conjecture how the social organisation of Western Europe, so favourable to the continued development of nations, has been brought about.

The great difference which divides the social organisation of the modern progressive peoples from that of all the ancient European civilisations is that, under it, the individual enjoys greater liberty and more securely founded rights as against the community, and as against all other individuals. This change is summed up in Sir H. Maine's dictum that "the movement of progressive societies has hitherto been a movement from status to contract."

All the ancient civilised societies, Greece and Rome no less than all the others, rested upon the fundamental assumption of the absolute supremacy of the State, the assumption that the individual existed only for the State and that the welfare of the State was the supreme end to which all individual rights and liberties must be subordinated absolutely, was the end to the securing of which all custom, and all law, all social and family relations and institutions, and religion itself were but the means. And the State was a politico-religious organisation, membership of which implied the blood relationship of its citizens and a common participation in the state religion; while the State gods were conceived as being themselves ancestors, or in some other way kinsmen, of the citizens.[1] This bond of blood or kinship between the members of the State and its gods went back to the earliest times. It is the rule of almost all savage peoples; and the religious rites of many include some rite symbolising or renewing this blood bond, such as smearing the blood of the kinsmen on the altars

[1] Cf. *La Cité Antique* of F. de Coulanges.

of the gods, or drinking the blood of some animal which is held to be the symbolic representative of the god. And the supreme end of the State itself was the increase of its own power and stability, through the exercise of military power and through military conquest.

All human beings outside the State, outside this moral-politico-religious bond, were regarded as *prima facie* enemies of the State, without rights of any sort, without even the slightest claim to humane treatment. Hence, in war the slaughter of the conquered was the rule; and the practice of making slaves of prisoners of war and of conquered peoples only arose through its profitableness, and was regarded as a great concession to the victims, whose natural fate was sudden death. Under this system, which inevitably became to some extent a caste system, with a caste of freemen or citizens ruling over slaves, each individual was born to a certain status as a member of a particular family. His position and duties and rights in the family were rigidly prescribed by custom, and the law took account only of the relations of the family to the State.

CHAPTER XX

The Progress of Nations in their Maturity

IN the foregoing chapter we have noted the great fact that the leading modern nations of Western Europe have shown a much greater capacity for progress than all the earlier civilised peoples, not excepting those of ancient Greece and Rome. I urged that this difference between the ancient and the modern European civilisations seemed to be chiefly due to a difference of social organisation. I pointed out how the older nations were essentially caste nations, resting on a basis of slavery, and how all individual rights were entirely subordinated to the welfare of the State, a politico-religious organisation held together by the bond of kinship; how, within that organisation, the rights of each person were strictly defined throughout his life by the status to which he was born; and how all persons outside this organisation were regarded as natural enemies, towards whom no obligations of any kind were felt. We have now to notice that the form of social organisation towards which all the leading modern nations have been tending, and which some of them have now pretty well achieved, is one in which the last vestiges of the caste system and the rigid bonds of customary status are rapidly being abolished. In this new organisation social classes persist, but they are no longer castes; all members of the nation are regarded as being by nature free and equal; a career is open to every talent, and any man may rise to any position by the exertion of his abilities. His position is one of extreme liberty as compared with that of any member of the ancient nations. He has definite rights as against the State. The State claims only a minimum of rights over him, the right to prevent him interfering with the rights of his fellow-citizens, the right to make him pay for his share of the privileges conveyed by its activities. And these rights it claims in virtue of contract between each citizen and all the rest. For each citizen is free to throw off his allegiance to the State and to leave it at will, and his continuance as a citizen of the State implies his acceptance of the contract.

Even in religion, personal liberty has at last been achieved; religion is no longer a State religion, the gods are no longer the national gods, and each man may accept any religion or none. This is the most striking instance of the immense distance, as regards the liberty of the individual, that divides the modern from the ancient nations. For with the latter, the function of religion was to preserve the security of the State; and to question it in any way was to threaten the State, a principle fully acted upon by Athens in the time of her highest enlightenment and glory.

The change is very striking also as regards the attitude of the citizens of one State towards those of any other and towards even the members of savage and barbarous communities. We no longer regard ourselves as devoid of all obligation towards such persons. Rather we tend to treat them as having equal rights with ourselves, the few specifically national rights excepted, and as having equal claims with our fellow citizens upon our considerate feeling and conduct towards them.

The relations of individuals are, then, tending to be regulated, on the one hand, by contractual justice; on the other hand, by the moral obligation felt by each individual—an obligation not enforced by any exercise of the power of the State, but supported only by public opinion. The end we set before ourselves is no longer the welfare of the State, to be attained at any cost to individual liberty; it is rather an ideal of justice for every person, to which the welfare of the State must be, if necessary, subordinated. In short, instead of maintaining universal intolerance, we have made great strides towards universal tolerance.

All this represents a profound change of social organisation, a great advance in social evolution. That it is intimately bound up with the progressiveness of a people, is shown by the fact that the degree to which the change has proceeded among the various nations runs parallel with their progress in all the essentials of civilisation. The change seems, indeed, to be one of the principal conditions of the progress of the nations of Western Europe and, we may add, of the American nation, by which it has been carried further than by any other. How, then, does it make for progress? We may answer this question by considering how the social system which has given place to this new kind of social organization—namely, the caste system—renders progress difficult or impossible.

Where the caste system is highly developed and rigidly maintained, as among the Hindus of India, its conservative unprogressive tendency is obvious enough. Each man is born a member of some one of many castes, and he can never hope to pass from one caste to another and higher caste. That fact alone removes at once the two greatest spurs to effort, the two most powerful motives that urge on the members of our modern societies to the fullest development and exercise of all their faculties; namely, the desire to rise in the social scale and to place one's children at a more advantageous starting point in the battle of life, and the fear of falling back in the social scale, of sinking to a lower level, with the consequent sacrifice of all the social consideration and other advantages which one's position at any given social level brings with it. Under the Hindu caste system, the poor Brahmin who has no possessions, perhaps not even a rag to cover his nakedness, is sure of the social consideration which his birth gives him, both for himself and his children. He can look disdainfully upon the rich man and the prince of lower caste; and public opinion approves and supports him. This perhaps is the most important way in which the caste system prevents progress. But there are others almost equally serious.

The occupations open to the members of each caste are rigidly limited. The members of one caste must be priests only, of another soldiers only, of another scavengers, of another potters, and so on. Now, if it were true that, when dexterities or mental powers generally are specially developed by use, the improvements of faculty resulting from this long practice and use were transmitted in any degree from generation to generation, we should expect the caste system to result, after many generations, in so many distinct breeds of men of highly specialised and perfected powers of the kinds used in the pursuit of each of the caste occupations. And this might make for progress. Each man would be employed in the occupation for which he was best suited. But, as we have seen, it is probable that use-inheritance does not occur; and there seems to be no evidence that differentiation and hereditary specialisation of faculties of this sort result from the caste system.[1] In each caste men continue to be born of the most diverse powers, suited for the most diverse occupations; and one effect of the caste system is that the best powers of any man will in the great majority of cases be prevented from finding their most effective outlet. That involves a great waste of faculty, which makes strongly for stagnation. We shall realise the importance of this influence, if we reflect on the great achievements, in the most

[1] A fact which provides another argument against use-inheritance.

diverse fields, of men who under our modern system have risen from humblest stations and occupations, to which under the caste system they would have been rigidly confined.

Again, within each caste custom rules the lives of the members with much greater force than it can exert in a large and complex society in the absence of the caste system. For each caste has its own tradition and customary code, which is necessarily narrow because of the uniformity of the conditions of life of those who obey it; hence tradition and custom have a narrow and well-defined field of operation; and the narrower the field of its application, the more rigidly will custom control action.

The caste system is thus one which permits of great differentiation and specialisation of pursuits, without any weakening of the conservative forces of society. It is for that reason presumably that the social organisation of all early civilisations tended to this form. It was the most easily attained form which combined diversity with stability sufficiently to permit of the formation of a large society or nation; and it was one which made for military efficiency. It formed, therefore, a natural stage of social evolution.

In so far as the caste system still survives, it owes its survival to the continuance of the need of the State for military efficiency. And we see how its maintenance is still rendered possible only by its alliance with a State religion and its system of religious sanctions. In Russia, for example, the caste system was thus maintained by the alliance of the military power with the religious system. While we see how in modern Germany the attempt to maintain the caste system and the supremacy of the State over the rights and liberties of individuals is breaking down, as the religious sanctions are losing their hold upon the people. Social democracy, secularism, and the demand for liberty go hand in hand.[1]

It is clear, then, that the caste system tends to produce a stable society and to prevent progress; and that, in proportion as it gives way to liberty and equality of all men, both legal and customary, and to the recognition of the rights of individuals as against the State, progress must be favoured.

In yet another way (perhaps more important than any other) the abolition of caste may favour development. In an earlier chapter I pointed out how every step in the development of the intellectual or the moral tradition of a people is initiated by some person of exceptional intellectual or moral power. I pointed out also how the existence of a hierarchy of social classes which are not exclusive castes, together with the operation of the social ladder by means of which individuals and families are enabled to climb up and down the social scale, tend to the segregation of ability in the classes of the upper part of the scale. They tend, in short, to produce classes capable of producing in each generation a relatively large number of persons of more than average capacities. Or, in other words, they lead to the concentration and mutual enrichment of the strains of exceptional capacity; they concentrate the best capacities of the people in a relatively small number of individuals of the favoured classes. And abilities so concentrated, and raised in a certain proportion of individuals to a higher power, will be more favourable in every way to the growth of the national mind than the same sum of abilities more evenly diffused throughout the population. At present it is impossible to say how far this segregation of abilities has gone, and what part it has played in forwarding the mental development of any na-

[1] This was written before the Great War but needs, I think, no modification.

tion. But that it has played some part, perhaps a very important one in some instances, can hardly be doubted.[1]

In Europe the feudal system served to tide over the period of transition from the ancient social organisation founded on caste and the supremacy of the politico-religious State to the modern system, the transition from the system founded on status and regulated by custom to the system founded upon equality and liberty and regulated by contract. For the feudal system, although still more or less a caste system, was nevertheless founded to some extent on contract. The tenure of land involved a contract to perform services in return; and such contract was the essence of the feudal system.

But it was not until the seventeenth and eighteenth centuries, when the feudal system was finally broken up and the Reformation initiated the age of religious tolerance, that the modern system substantially replaced the ancient system, and the age of rapid progress set in.

In proportion as the change was achieved, the powers of all men were set free in an unfettered competition such as had never before been possible. Independence of thought and action, free discussion of all principles, and the recognition of the relativity of all truths, succeeded to unquestioning subservience to ancient formulas and customs. Each man became comparatively free to follow his natural bent, to develop his best powers to the utmost, and to secure by the exercise of those powers the maximum of social consideration and of well-being, unfettered by arbitrary restrictions, civil or ecclesiastical. I think we may fairly say that the modern pragmatic or humanistic

[1] Francis Galton and his disciples have produced much evidence to show that the educated class of Englishmen includes a very much larger proportion of strains of high ability than the rest of the people, it having been formed by the long continued operation of the social ladder. There is no reason to doubt the truth of this conclusion.

movement in philosophy, in the midst of which we are living, represents the final stage of this emancipation of man from the bonds which he has created for himself.

We have seen, then, that the modern system of social organisation does not make for the racial progress of a whole people, but probably up to the present time, for race-deterioration; nevertheless, it certainly makes for progress of the intellectual and moral traditions of peoples; and we can now see in what way it makes for progress. The improvement of racial qualities by natural selection of the innately superior individuals has been brought to an end; the mortal conflict of societies has also practically been abolished as a factor of race progress, as also of collective or social progress. These have been replaced by a new form of struggle for existence and of selection—namely, the rivalry and competition of ideas and of the institutions in which ideas become embodied, and the selection for survival of those ideas and institutions which are found, under the tests of practice and experience, most accordant with the truth and, therefore, best adapted to promote the welfare of societies and of their members.[1]

And this process of survival of the fittest and elimination of the unfit among ideas and institutions takes place not only within nations, but has also international scope. The members of each nation no longer, as of old, regard all foreigners as their natural enemies, no longer despise their institutions and reject their ideas with scorn. They are ready to learn from others, to let the ideas current among other peoples enter into competition with native-born ideas; and so the number and variety of competing elements increases, and the intensity of the competition waxes ever keener. Every idea that constitutes an important advance in our intellectual outlook or in our practical

[1] Prof. S. Alexander, in his *Moral Order and Progress*, was perhaps the first to draw attention to this form of the struggle for existence.

command over nature rapidly finds acceptance throughout the civilised world and displaces some less true idea, some less appropriate institution, some less effective mode of action.

Two great conditions, making for continued improvement of the moral and intellectual traditions, characterise, then, Western civilisation. First, within each nation there is going on the process of emancipation of all human faculties, so that they enter into the freest possible competition with one another on a footing of equality; this process, although now far advanced in all the leading nations, is still being carried further, and the whole trend of modern legislation is to confirm the change and hasten it to its completion.[1] Secondly, there is a circle of peoples whose ideas are thrown into the arena of rivalry, to suffer extinction or to gain universal acceptance. This circle also is constantly widening by the inclusion of peoples hitherto outside it; and each new admission, as of Japan in recent years, is a new stimulus to the further evolution of the collective mind of each nation concerned. Both these conditions depend upon improved social organisation.

How then has this great change of social organisation been effected? To put this question is to approach an immense subject, the history of liberty and toleration. I can only make one or two brief remarks. It has been suggested by many authors, notably by Kidd in his volume on *Western Civilisation*, that we owe this great change to the Christian religion. It is pointed out that the Christian religion, unlike most earlier religions, was from the first not a national or State religion but a universal religion, and that its adoption has weakened the tyranny of the State by breaking up its alliance with religion. Further, it is a religion which, by its doctrine of the immortal-

[1] This last sentence perhaps is only partially true. A rigid system of State Socialism would involve a retrogression in this respect.

ity of the souls of all men, has tended to give dignity and value to each individual life, quite independently of personal status. Again, by its teaching of universal charity, it has to some extent softened and moralised the relations of men and of societies. But, that the replacement of a national religion by a universal religion which teaches the equality of all believers does not suffice to secure continued collective evolution is shown by the instances of Buddhism and Mohammedanism. Both of these are of this character; yet both have failed to render continuously progressive the societies that have accepted them.

That the spread of the Christian religion does not in itself suffice to account for the evolution from the ancient to the modern type of social organisation is shown also by the fact that it had held undisputed sway among the peoples of Western Europe for more than a thousand years before social evolution made any considerable advance. Throughout that period, religion constantly called in the civil and military power of the State to enforce the acceptance of its dogmas. And that its teachings did not suffice to produce religious or civil tolerance is shown by the fierce and incessant persecutions of heretics and the many religious wars that fill the history of medieval Europe.

The religious tolerance and liberty of the modern era are rather features of a wider phenomenon, the general increase of tolerance and liberty; and they must be ascribed to the same causes as this wider fact. They imply a great evolution of the moral tradition, the most important and striking feature of which is the expansion of the sphere in which the sympathetic feelings find application. There is no reason to suppose that the feelings and emotions underlying the sympathetic and considerate treatment of others have changed in character in the historic period. For long ages men have felt such sympathy and given considerate and just treatment to those who have been nearest to them; at first to the members of their own immediate family; later to the fellow-members of their own small society; and then, as societies expanded into complex caste societies, to the members of their own caste; later, as castes were broken down, to all their fellow citizens; and still later in some degree to all men.

It is this progressive extension of the sphere of imaginative sympathy which, more than anything else, has broken down all the social barriers that confined the energies of men and has set free their various faculties in that competition of ever increasing severity which is the principal cause underlying the modern progress of peoples. It is this which has destroyed nearly all the old bonds that fettered and limited men's activities in religion, in science, in politics, in art, in commerce, in manufacture, and has brought men in all these spheres into that intense, because free and equal, competition, which produces an ever accelerating progress. It is this which has produced the almost universal acceptance of the entirely and most characteristically modern principle of "one man one vote," a principle so hard to justify on any ground of expediency, from any considerations of the stability and welfare of the State. It is this also which has led to so greatly increased intercourse between peoples.

It is sometimes contended that the realisation of the principles of equality and justice for all men has been secured only by the strife of the social classes, by the success of the lower classes in forcing a series of concessions from the ruling classes. This is a very imperfect and partial view of the process. If the ruling classes had consistently sought to maintain their power and exclusive privileges, and to maintain all the rest of society in a state of servitude or serfdom, there is little doubt that they could have done so. But their position has been weakened from within by the extension of their sympathies. Consider the great series of legislative changes which, during the nineteenth century, transformed the social organisation of this country, especially the factory laws, the franchise extension laws, and the laws for the abolition of slavery. These were for the most part of the nature of a voluntary abdication of power on the part of the classes in possession. Consider the topics which chiefly engross the attention of our legislators and are the centre of political and social discussion. They are the providing of a better and freer education for the children of the working classes, who of themselves would probably never have thought of such a thing; the providing of free meals for school children; the providing of work and food for the unemployed; temperance laws, land settlement, and emigration, the eight-hours day, housing of the working classes, free trade and cheap food, old age pensions; all measures for raising the standard of life of the labouring classes and securing them against the tyranny of capital.

In respect of our relations to the lower peoples the same proposition holds good. It would be easy for the European nations to exterminate the black people of Africa, and to possess themselves of all their lands.[1] But public opinion will not now allow this; it insists upon our moral obligation towards such peoples, that we are bound to try to help them to survive and to raise themselves to our level of culture.

The extension of the sphere of application of imaginative sympathy has, then, been a factor of prime importance in producing the social evolution which underlies modern progress.

The factors that have brought about this extension have been many and complex, and it is perhaps a hopeless

[1] As the Spaniard well-nigh exterminated in the name of the Church the civilisation and the nations of Mexico and Peru.

task to attempt to enumerate them and to apportion to each its share of influence. Undoubtedly, it has been produced largely by the influence of a relatively small number of enlightened leaders of opinion, such men as Wilberforce, Stuart Mill, Shaftesbury, John Locke, Rousseau, and Voltaire—men whose original intellectual powers enabled them to criticise and reject the settled principles of their time. It was a work of liberation from custom and traditional prejudice effected by the spirit of inquiry, which questioned the validity of the old narrow conceptions of the relations of men and peoples, the old narrow prejudices of caste and nation, and discovered their fallacies to the world; discovered, for example, that men of a religious persuasion slightly different from one's own are not necessarily wicked, nor those of a different nationality necessarily despicable and possessed of no ideas worthy of admiration and adoption.

But the ground was prepared for the reception of the teachings of such men by the conflicts of men who desired nothing of tolerance and equality and liberty. This is best illustrated by the history of religious toleration. As I said before, religion is essentially conservative and intolerant of heresies. The first effect on religion of that revival and liberation of the spirit of inquiry which we call the Renascence, was to produce not religious toleration, but rather a bitter conflict of mutually intolerant sects. And religious toleration was eventually achieved largely by the realisation of the necessity of compromise among these warring and constantly multiplying sects; it was found impossible to weed out heresy by persecution. Yet who can doubt that the Church, if it believed that it saw its way to secure the universal acceptance of its doctrines by means of persecution, would long hesitate to return to its ancient practices? The coming of religious toleration was due to the application of the spirit of inquiry to religious systems; these inquiries produced irreconcilable sects, whose strife prepared the way for compromise and toleration.

The strife of parties and sects was itself part of a still wider process; and this process must be recognised as the most important single condition of that widening of the sphere of imaginative sympathy which has been the root cause of the improvement of social organisation, of the general increase of liberty, and thus of the progress of the modern nations. This wider process is the general increase of human intercourse, both within nations and between them. Only so long as men know little of one another, can they continue to regard one another with entire hostility or cold indifference. The knowledge and understanding brought by personal intercourse is necessary to sympathy; but as soon as, and in proportion as, such knowledge is acquired, the innate social tendencies common to all men are brought into play. As soon as man understands that his fellow man suffers the same pains and joys as himself, longs for the same goods, fears the same evils, throbs with the same emotions and desires, then he shares with him in some degree these feelings, in virtue of that fundamental law of all social beings, the law of primitive sympathy; then also pity and sympathetic sorrow and tender regard are awakened in his breast; then his fellow man is no longer the object of his cold or hostile glances, as a certain rival and probable enemy, but is seen to be a fellow toiler and sufferer whom he is willing to succour, a fellow creature whose joys and sorrows alike he cannot but share in some degree.

Increasing freedom of intercourse throughout the civilised world, and beyond its boundaries also, has been the most characteristic feature of the age of progress; and in it we may recognise the most fundamental condition of that progress. Science and mechanical invention have been the means by which this greater freedom of intercourse has been brought about. First and most important perhaps was the invention of printing, the consequent spread of the habit of reading, and the wide diffusion of the written word. Second only to this was the improvement of the art of navigation, which brought the remotest peoples of the world within the ken of Europe and greatly promoted the intercourse of the European peoples, as well as the circulation of persons and news within each nation; for the development of commerce over seas implies a corresponding development of commerce within the national boundaries. Then came the use of steam in locomotion on sea and land, the press and the telegraph; and, with the advent of these, intercourse within and without became really free and abundant; mutual knowledge and understanding between men and nations grew rapidly, and the age of progress was assured.

The progressive character of the modern nations has been due, then, to the actions and reactions between the spirit of inquiry and the improvement of forms of social organisation; each step in the one respect has reacted upon the other, stimulating further change in the same direction. And the medium through which they have chiefly thus worked upon one another has been the increase of intercourse between men and nations. The spirit of inquiry has urged men on to explore their fellow men and to study foreign nations, and it has provided the means for so doing; the greater mutual knowledge and sympathy thus brought into being have in turn brought greater liberty to the spirit of inquiry, freeing it from the rigid bonds of custom and conservative tradition and enabling it to render human intercourse yet more free and abundant.

In this way we reconcile and synthesise the rival theories of the causes of progress, the view that sees in the spirit of inquiry the sole agent of progress and that which attributes it wholly to the improvement of morals and of social organisation. The great commandment, "Thou shalt love thy neighbour as thyself," pointed the way of all progress; but great and beautiful as it was, it could not immediately avail to break the bonds of the human spirit, the bonds of ignorance and fear; only gradually through increase of knowledge could man learn that all men are his neighbours, and that not only the foreigner just beyond the frontier, but also the naked savage, chipping his stone axe or weaving his rude basket for the reception of his neighbour's head, is a man of like passions with himself, with equal claims upon justice and freedom and all that makes the humanity of man.

It only remains to point out the part in human life of a new factor of progress which promises to eclipse all others in importance. The main theme of my earlier work[1] was that only through increase of knowledge of others is each man's knowledge of himself slowly built up and enriched, until it renders him capable of enlightened self-direction. So the main theme of this book is the development of the group mind, the increase of its self-knowledge and of its power of self-direction through increase of knowledge of other human societies.

The age of progress through which the world has recently passed was an age of progress due to increase of human intercourse and consequent increasing understanding by each nation of other nations and peoples. This better knowledge of other peoples is now reacting upon the self-knowledge of each nation, rapidly enriching it. Each of the great nations is beginning to understand itself, and to take thought for the morrow in the light of this self-knowledge; and this increase of national self-knowledge, this enrichment of national self-consciousness, is the great new factor which alone can secure the further

[1] *Introduction to Social Psychology.*

progress of mankind. We saw in an earlier chapter that a nation is essentially the realisation of an idea, the idea of a nation; that only in so far as the idea of the nation exists and operates in the minds of the members of the nation, controlling their conduct and directing it to actions having reference to the nation as a whole, does a nation come into and continue in existence. The self-consciousness of nations is therefore not a new factor in their life. But their self-consciousness is now becoming reflective and immensely richer in content; so much so that it promises to operate virtually as a new factor of tremendous efficiency.

We may illustrate the influence of this new factor by reverting again to the analogy between the mind of the individual and the mind of the nation which we developed at some length in an earlier chapter. In the developing individual, as in the evolving animal series, the development of self-consciousness is the condition of the development of true volition. Before self-consciousness and a self-regarding sentiment are developed, conduct is determined by feelings and impulses, or by ideas and the desires they arouse, either some one desire rising alone to consciousness and issuing at once in action, or through a conflict of impulses and desires, some one of which eventually predominates over the others and determines action; but action issuing from such a conflict of impulses and desires is not true volition. Action is truly volitional only when the ideal of the self in relation to the idea of the end to be achieved by each of the conflicting tendencies determines the issue of the conflict.

In the mental life of nations, all those conflicts of ideas, of parties, of principles, and of systems, in which each strives to predominate over and displace others, and by natural selection of which (the death of the many less fit, the survival of the few better or more fit) the progress of recent centuries has been chiefly due; all these conflicts have been more or less blind conflicts, in which the idea of the whole nation, in relation to the end to be achieved by each of the conflicting tendencies, has generally played but a small part and a part that often has not made strongly for progress. National actions were in the main impulsive and instinctive actions, like those of young children or the higher animals. And for this reason—that nations had too little true self-knowledge, and had not developed a true and rich ideal of national life—the self-consciousness of nations was too poor in content to serve as the guide of actions making for progress.

In the individual man, it is the growing richness and accuracy of self-knowledge which alone enables him to direct his actions effectively to secure his own welfare and to improve his character and powers. Just so in nations the rapid growth of their self-knowledge and the enrichment of their ideals of national life which characterise the present time must render their self-consciousness a far more efficient guide of all national deliberation and action.

The self-knowledge of the individual grows chiefly, as we have seen, through intercourse with his fellows; his idea of himself develops in fulness and accuracy in the light of his knowledge of other selves, and this knowledge in turn develops in the light of his increasing knowledge of himself. Just so the self-knowledge of nations is now growing rapidly through the intercourse of each nation with others, an intercourse far freer, more multiplex, than ever before in the history of the world; a result largely of the improved means of communication which we owe to science and the spirit of inquiry.

Perhaps the most striking illustration of the operation of this new factor is the rapid spread in recent years of parliamentary institutions. The parliamentary system of national organisation was worked out in these islands by long centuries of more or less blind conflict of ideas and parties and institutions; and now other nations in rapid succession have observed and admired the system and have deliberately and self-consciously adopted it; and still the process goes on, as recently in Russia.

Japan offers a striking illustration of the way in which the new factor operates. An intelligent people in which the national sentiment was strong, but in which national self-knowledge was rudimentary because of the isolation of the nation, was suddenly brought into contact with other peoples; through observation of them, it learnt its own deficiencies and set about deliberately to remedy them in the light of its new knowledge; and in doing so has reorganised itself from top to bottom.

In England also national self-knowledge is beginning rapidly to increase in accuracy and extent. We have begun to compare ourselves at all points with other nations, and are no longer content with the good old creed, that everything British is best. We are learning in this way our weaknesses; and the knowledge is becoming a main cause of accelerated progress. The best illustration is, perhaps, the present stir over educational questions, which is directly due to the increase of national self-knowledge resulting from the observation of other nations.

But in the future our national self-consciousness will be enriched and fitted for the guidance of the national will in a still more effective manner than by the knowledge of our weaknesses being forced upon us by the nations who are our rivals in the world. In many directions—by the historians, the biologists, the anthropologists, the statisticians—data are being gathered for a science of society whose sure indications will enable us deliberately to guide the further evolution of the nation towards the highest ideal of a nation that we can conceive. In this way, it may be hoped, the modern nations will be able to avoid that danger which has destroyed the great nations of the past, and which has been the dark cloud shadowing the brilliance of the age of progress that resulted from increasing human intercourse and mutual understanding. In this way the free play of the spirit of inquiry, which in all earlier ages has been highly dangerous to the stability of nations and which, while it was the sole cause of progress, nevertheless destroyed many of the nations whom it impelled upon that path, will make for a greatly accelerated progress; and, at the same time, it will enable us to secure, by deliberate voluntary control, the bases of society, which in all previous ages have rested solely upon custom, instinct, and the religious sanctions.

Not by any voluntary surrender of the reason, not by any subjugation of the intellect to the dominion of obscure transcendental ideas, such as is preached by Benjamin Kidd, Chatterton-Hill,[1] and others who have realised the disintegrating effects of intellect on earlier societies, but by a more strenuous use of our intellectual faculties, and by a growth of knowledge, especially a knowledge of the laws of human societies, will the stability and further evolution of nations be maintained.

The nations whose progress will rest upon this basis will be in a position very different from that of the older societies to which the emancipation of the intellect was fatal. They fell for lack of knowledge of natural laws, as soon as the progress of intellectual inquiry had weakened their instinctive and customary bases. The modern nations may reasonably hope that they are within sight of knowledge which will enable them to avoid these dangers and to continue their progress during an indefinitely long period. They may even hope to progress, not only in respect of the intellectual and moral tradition, but also in

[1] *Heredity and Selection in Sociology*, London, 1907: an interesting work similar in tendency to Kidd's *Social Evolution*.

respect of racial qualities; for a better knowledge of the factors at work and of the laws of heredity will enable them to put an end to the influences now making for race deterioration and to replace them by others of the opposite tendency.

Such national progress will be truly teleological; it will be a progress whose direction will be determined by the desire of an ideal end, present to the consciousness of all and striven after by the collective deliberation and volition of the nation.

Thus the group spirit, rising above the level of a narrow patriotism that regards with hostility all its rivals, recognising that only through the further development of the collective life of nations can man rise to higher levels than he has yet known, becomes the supreme agent of human progress.

INDEX